Geographic Information Science
and Systems

Geographic Information Science and Systems

Editor: Augustus Winstead

NY RESEARCH P R E S S

New York

Published by NY Research Press
118-35 Queens Blvd., Suite 400,
Forest Hills, NY 11375, USA
www.nyresearchpress.com

Geographic Information Science and Systems
Edited by Augustus Winstead

International Standard Book Number: 978-1-63238-595-6 (Hardback)

Cataloging-in-Publication Data

Geographic information science and systems / edited by Augustus Winstead.
   p. cm.
Includes bibliographical references and index.
ISBN 978-1-63238-595-6
1. Geographic information systems. 2. Information storage and retrieval systems--Geography.
I. Winstead, Augustus

G70.212 .G46 2018
910.285--dc23

Contents

Preface

Geographic information system refers to the system which is used to present, design, manage, store, capture, control, analyze and manipulate geographic and spatial data. As a sub-field of geoinformatics, GIS concerns itself with the academic study of the geographic data and the methods used to calculate and record it. Geographic information systems play a significant role in hydrological, topological and cartographic modeling. The topics included in this book on GIS are of utmost significance and bound to provide incredible insights to readers. This textbook is an essential guide for both academicians and those who wish to pursue this discipline further.

To facilitate a deeper understanding of the contents of this book a short introduction of every chapter is written below:

Chapter 1- The system used for capturing, storing and studying geographical data is known as geographical information system. Some of the fields where GIS can be used include public health, crime mapping, landscape architecture, national defense and archaeology. Geographical information system is an emerging field of study; the following chapter will not only provide an overview, it will also delve deep into the variegated topics related to it.

Chapter 2- Maps are a visual representation of objects and regions that are found on Earth. Maps can be divided into scale based maps and purpose scale maps. Map scaling has improved with computer aided designs and graphics. Maps and map scaling can best be understood in confluence with the major topics listed in the following section.

Chapter 3- The science of making maps is known as cartography. With the advancement of technology, printing press and vernier has made mass production of maps possible. Cartography can be divided into two categories, thematic cartography and general cartography. The chapter serves as a source to understand the major categories related to cartography.

Chapter 4- Spatial database is the database that stores and retrieves data that defines a geometric space. It represents simple objects like lines or polygons and also complex structures like topological coverages and Triangulated irregular networks (TINs). This section discusses the methods of GIS database and spatial analysis in a critical manner providing key analysis to the subject matter.

I owe the completion of this book to the never-ending support of my family, who supported me throughout the project.

Editor

An Overview of GIS

The system used for capturing, storing and studying geographical data is known as geographical information system. Some of the fields where GIS can be used include public health, crime mapping, landscape architecture, national defense and archaeology. Geographical information system is an emerging field of study; the following chapter will not only provide an overview, it will also delve deep into the variegated topics related to it.

Geographic Information System (GIS)

GIS stands for Geographical Information System. It is defined as an integrated tool, capable of mapping, analyzing, manipulating and storing geographical data in order to provide solutions to real world problems and help in planning for the future. GIS deals with what and where components of occurrences. For example, to regulate rapid transportation, government decides to build fly-over (what component) in those areas of the city where traffic jams are common (where component).

GIS means differently to different people and therefore has different definitions. For example, Burrough (1998) defined GIS as " a powerful set of tools for collecting, storing, retrieving at will, transforming and displaying spatial data from the real world for a particular set of purposes"

Objectives of GIS

Some of the major objectives of GIS are to:

- Maximizing the efficiency of planning and decision making
- Integrating information from multiple sources
- Facilitating complex querying and analysis
- Eliminating redundant data and minimizing duplication

Components of a GIS

A GIS has following components:

Hardware: It consists of the equipments and support devices that are required to capture, store process and visualize the geographic information. These include computer with hard disk, digitizers, scanners, printers and plottersetc.

Software: Software is at the heart of a GIS system. The GIS software must have the basic capabilities of data input, storage, transfosrmation, analysis and providing desired outputs. The interfaces could be different for different softwares. The GIS softwares being used today belong to either of the category –proprietary or open source. ArcGIS by ESRI is the widely used proprietary GIS software. Others in the same category are MapInfo, Microstation, Geomedia etc. The development of open source GIS has provided us with freely available desktop GIS such as Quantum, uDIG, GRASS, MapWindow GIS etc., GIS softwares.

Data: The data is captured or collected from various sources (such as maps, field observations, photography, satellite imagery etc) and is processed for analysis and presentation.

Procedures: These include the methods or ways by which data has to be input in the system, retrieved, processed, transformed and presented.

People: This component of GIS includes all those individuals (such as programmer, database manager, GIS researcher etc.) who are making the GIS work, and also the individuals who are at the user end using the GIS services, applications and tools.

A geographic information system (GIS) is a system designed to capture, store, manipulate, analyze, manage, and present spatial or geographic data. The acronym GIS is sometimes used for geographic information science (GIScience) to refer to the academic discipline that studies geographic information systems and is a large domain within the broader academic discipline of geoinformatics. What goes beyond a GIS is a spatial data infrastructure, a concept that has no such restrictive boundaries.

In general, the term describes any information system that integrates, stores, edits, analyzes, shares, and displays geographic information. GIS applications are tools that allow users to create interactive queries (user-created searches), analyze spatial information, edit data in maps, and present the results of all these operations. Geographic information science is the science underlying geographic concepts, applications, and systems.

GIS is a broad term that can refer to a number of different technologies, processes, and methods. It is attached to many operations and has many applications related to engineering, planning, management, transport/logistics, insurance, telecommunications, and business. For that reason, GIS and location intelligence applications can be the foundation for many location-enabled services that rely on analysis and visualization.

GIS can relate unrelated information by using location as the key index variable. Locations or extents in the Earth space–time may be recorded as dates/times of occurrence, and x, y, and z coordinates representing, longitude, latitude, and elevation, respectively.

All Earth-based spatial–temporal location and extent references should be relatable to one another and ultimately to a "real" physical location or extent. This key characteristic of GIS has begun to open new avenues of scientific inquiry.

GIS Techniques and Technology

Modern GIS technologies use digital information, for which various digitized data creation methods are used. The most common method of data creation is digitization, where a hard copy map or survey plan is transferred into a digital medium through the use of a CAD program, and geo-referencing capabilities. With the wide availability of ortho-rectified imagery (from satellites, aircraft, Helikites and UAVs), heads-up digitizing is becoming the main avenue through which geographic data is extracted. Heads-up digitizing involves the tracing of geographic data directly on top of the aerial imagery instead of by the traditional method of tracing the geographic form on a separate digitizing tablet (heads-down digitizing).

Relating Information from Different Sources

GIS uses spatio-temporal (space-time) location as the key index variable for all other information. Just as a relational database containing text or numbers can relate many different tables using common key index variables, GIS can relate otherwise unrelated information by using location as the key index variable. The key is the location and/or extent in space-time.

Any variable that can be located spatially, and increasingly also temporally, can be referenced using a GIS. Locations or extents in Earth space–time may be recorded as dates/times of occurrence, and x, y, and z coordinates representing, longitude, latitude, and elevation, respectively. These GIS coordinates may represent other quantified systems of temporo-spatial reference (for example, film frame number, stream gage station, highway mile-marker, surveyor benchmark, building address, street intersection, entrance gate, water depth sounding, POS or CAD drawing origin/units). Units applied to recorded temporal-spatial data can vary widely (even when using exactly the same data), but all Earth-based spatial–temporal location and extent references should, ideally, be relatable to one another and ultimately to a "real" physical location or extent in space–time.

Related by accurate spatial information, an incredible variety of real-world and projected past or future data can be analyzed, interpreted and represented. This key characteristic of GIS has begun to open new avenues of scientific inquiry into behaviors and patterns of real-world information that previously had not been systematically correlated.

GIS Uncertainties

GIS accuracy depends upon source data, and how it is encoded to be data referenced. Land surveyors have been able to provide a high level of positional accuracy utilizing the GPS-derived positions. High-resolution digital terrain and aerial imagery, powerful computers and Web technology are changing the quality, utility, and expectations of GIS to serve society on a grand scale, but nevertheless there are other source data that affect overall GIS accuracy like paper maps, though these may be of limited use in achieving the desired accuracy.

In developing a digital topographic database for a GIS, topographical maps are the main source, and aerial photography and satellite imagery are extra sources for collecting data and identifying attributes which can be mapped in layers over a location facsimile of scale. The scale of a map and geographical rendering area representation type are very important aspects since the information content depends mainly on the scale set and resulting locatability of the map's representations. In order to digitize a map, the map has to be checked within theoretical dimensions, then scanned into a raster format, and resulting raster data has to be given a theoretical dimension by a rubber sheeting/warping technology process.

A quantitative analysis of maps brings accuracy issues into focus. The electronic and other equipment used to make measurements for GIS is far more precise than the machines of conventional map analysis. All geographical data are inherently inaccurate, and these inaccuracies will propagate through GIS operations in ways that are difficult to predict.

Data Representation

GIS data represents real objects (such as roads, land use, elevation, trees, waterways, etc.) with digital data determining the mix. Real objects can be divided into two abstractions: discrete objects (e.g., a house) and continuous fields (such as rainfall amount, or elevations). Traditionally, there are two broad methods used to store data in a GIS for both kinds of abstractions mapping references: raster images and vector. Points, lines, and polygons are the stuff of mapped location attribute references. A new hybrid method of storing data is that of identifying point clouds, which combine three-dimensional points with RGB information at each point, returning a "3D color image". GIS thematic maps then are becoming more and more realistically visually descriptive of what they set out to show or determine.

Data Capture

Data capture—entering information into the system—consumes much of the time of GIS practitioners. There are a variety of methods used to enter data into a GIS where it is stored in a digital format.

Existing data printed on paper or PET film maps can be digitized or scanned to produce digital data. A digitizer produces vector data as an operator traces points, lines, and polygon boundaries from a map. Scanning a map results in raster data that could be further processed to produce vector data.

The image shows example of hardware for mapping (GPS and laser rangefinder) and data collection (rugged computer). The current trend for geographical information system (GIS) is that accurate mapping and data analysis are completed while in the field. Depicted hardware (field-map technology) is used mainly for forest inventories, monitoring and mapping.

Survey data can be directly entered into a GIS from digital data collection systems on survey instruments using a technique called coordinate geometry (COGO). Positions from a global navigation satellite system (GNSS) like Global Positioning System can also be collected and then imported into a GIS. A current trend in data collection gives users the ability to utilize field computers with the ability to edit live data using wireless connections or disconnected editing sessions. This has been enhanced by the availability of low-cost mapping-grade GPS units with decimeter accuracy in real time. This eliminates the need to post process, import, and update the data in the office after fieldwork has been collected. This includes the ability to incorporate positions collected using a laser rangefinder. New technologies also allow users to create maps as well as analysis directly in the field, making projects more efficient and mapping more accurate.

Remotely sensed data also plays an important role in data collection and consist of sensors attached to a platform. Sensors include cameras, digital scanners and lidar, while platforms usually consist of aircraft and satellites. In England in the mid 1990s, hybrid kite/balloons called Helikites first pioneered the use of compact airborne digital cameras as airborne Geo-Information Systems. Aircraft measurement software, accurate to 0.4 mm was used to link the photographs and measure the ground. Helikites are inexpensive and gather more accurate data than aircraft. Helikites can be used over roads, railways and towns where UAVs are banned.

Recently with the development of miniature UAVs, aerial data collection is becoming possible with them. For example, the Aeryon Scout was used to map a 50-acre area with a Ground sample distance of 1 inch (2.54 cm) in only 12 minutes.

The majority of digital data currently comes from photo interpretation of aerial photographs. Soft-copy workstations are used to digitize features directly from stereo pairs of digital photographs. These systems allow data to be captured in two and three dimensions, with elevations measured directly from a stereo pair using principles of photogrammetry. Analog aerial photos must be scanned before being entered into a soft-copy system, for high-quality digital cameras this step is skipped.

Satellite remote sensing provides another important source of spatial data. Here satellites use different sensor packages to passively measure the reflectance from parts of the electromagnetic spectrum or radio waves that were sent out from an active sensor such as radar. Remote sensing collects raster data that can be further processed using different bands to identify objects and classes of interest, such as land cover.

When data is captured, the user should consider if the data should be captured with either a relative accuracy or absolute accuracy, since this could not only influence how information will be interpreted but also the cost of data capture.

After entering data into a GIS, the data usually requires editing, to remove errors, or further processing. For vector data it must be made "topologically correct" before it can be used for some advanced analysis. For example, in a road network, lines must connect with nodes at an intersection. Errors such as undershoots and overshoots must also be removed. For scanned maps, blemishes on the source map may need to be removed from the resulting raster. For example, a fleck of dirt might connect two lines that should not be connected.

Raster-to-vector Translation

Data restructuring can be performed by a GIS to convert data into different formats. For example, a GIS may be used to convert a satellite image map to a vector structure by generating lines around all cells with the same classification, while determining the cell spatial relationships, such as adjacency or inclusion.

More advanced data processing can occur with image processing, a technique developed in the late 1960s by NASA and the private sector to provide contrast enhancement, false color rendering and a variety of other techniques including use of two dimensional Fourier transforms. Since digital data is collected and stored in various ways, the two data sources may not be entirely compatible. So a GIS must be able to convert geographic data from one structure to another. In so doing, the implicit assumptions behind different ontologies and classifications require analysis. Object ontologies have gained increasing prominence as a consequence of object-oriented programming and sustained work by Barry Smith and co-workers.

Projections, Coordinate Systems, and Registration

The earth can be represented by various models, each of which may provide a different set of coordinates (e.g., latitude, longitude, elevation) for any given point on the Earth's surface. The simplest model is to assume the earth is a perfect sphere. As more measurements of the earth have accumulated, the models of the earth have become more sophisticated and more accurate. In fact, there are models called datums that apply to different areas of the earth to provide increased accuracy, like NAD83 for U.S. measurements, and the World Geodetic System for worldwide measurements.

Spatial Analysis with Geographical Information System (GIS)

GIS spatial analysis is a rapidly changing field, and GIS packages are increasingly including analytical tools as standard built-in facilities, as optional toolsets, as add-ins or 'analysts'. In many instances these are provided by the original software suppliers (commercial vendors or collaborative non commercial development teams), while in other cases facilities have been developed and are provided by third parties. Furthermore, many products offer software development kits (SDKs), programming languages and language support, scripting facilities and/or special interfaces for developing one's own analytical tools or variants. The website "Geospatial Analysis" and associated book/ebook attempt to provide a reasonably comprehensive guide to the subject. The increased availability has created a new dimension to business intelligence termed "spatial intelligence" which, when openly delivered via intranet, democratizes access to geographic and social network data. Geospatial intelligence, based on GIS spatial analysis, has also become a key element for security. GIS as a whole can be described as conversion to a vectorial representation or to any other digitisation process.

Slope and Aspect

Slope can be defined as the steepness or gradient of a unit of terrain, usually measured as an angle in degrees or as a percentage. Aspect can be defined as the direction in which a unit of terrain faces. Aspect is usually expressed in degrees from north. Slope, aspect, and surface curvature in terrain analysis are all derived from neighborhood operations using elevation values of a cell's adjacent neighbours. Slope is a function of resolution, and the spatial resolution used to calculate slope and aspect should always be specified. Authors such as Skidmore, Jones and Zhou and Liu have compared techniques for calculating slope and aspect.

The following method can be used to derive slope and aspect:

The elevation at a point or unit of terrain will have perpendicular tangents (slope) passing through the point, in an east-west and north-south direction. These two tangents give two components, $\partial z/\partial x$ and $\partial z/\partial y$, which then be used to determine the overall direction of slope, and the aspect of the slope. The gradient is defined as a vector quantity with components equal to the partial derivatives of the surface in the x and y directions.

The calculation of the overall 3x3 grid slope S and aspect A for methods that determine east-west and north-south component use the following formulas respectively:

$$\tan S = \sqrt{\left(\frac{\partial z}{\partial x}\right)^2 + \left(\frac{\partial z}{\partial y}\right)^2}$$

$$\tan A = \left(\frac{\left(\frac{-\partial z}{\partial y}\right)}{\left(\frac{\partial z}{\partial x}\right)}\right)$$

Zhou and Liu describe another formula for calculating aspect, as follows:

$$A = 270^\circ + \arctan\left(\frac{\left(\frac{\partial z}{\partial x}\right)}{\left(\frac{\partial z}{\partial y}\right)}\right) - 90^\circ \left(\frac{\left(\frac{\partial z}{\partial y}\right)}{\left|\frac{\partial z}{\partial y}\right|}\right)$$

Data Analysis

It is difficult to relate wetlands maps to rainfall amounts recorded at different points such as airports, television stations, and schools. A GIS, however, can be used to depict two- and three-dimensional characteristics of the Earth's surface, subsurface, and atmosphere from information points. For example, a GIS can quickly generate a map with isopleth or contour lines that indicate differing amounts of rainfall. Such a map can be thought of as a rainfall contour map. Many sophisticated methods can estimate the characteristics of surfaces from a limited number of point measurements. A two-dimensional contour map created from the surface modeling of rainfall point measurements may be overlaid and analyzed with any other map in a GIS covering the same area. This GIS derived map can then provide additional information - such as the viability of water power potential as a renewable energy source. Similarly, GIS can be used to compare other renewable energy resources to find the best geographic potential for a region.

Additionally, from a series of three-dimensional points, or digital elevation model, isopleth lines representing elevation contours can be generated, along with slope analysis, shaded relief, and other elevation products. Watersheds can be easily defined for any given reach, by computing all of the areas contiguous and uphill from any given point of interest. Similarly, an expected thalweg of where surface water would want to travel in intermittent and permanent streams can be computed from elevation data in the GIS.

Topological Modeling

A GIS can recognize and analyze the spatial relationships that exist within digitally stored spatial data. These topological relationships allow complex spatial modelling and analysis to be performed. Topological relationships between geometric entities traditionally include adjacency (what adjoins what), containment (what encloses what), and proximity (how close something is to something else).

Geometric Networks

Geometric networks are linear networks of objects that can be used to represent interconnected features, and to perform special spatial analysis on them. A geometric network is composed of edges, which are connected at junction points, similar to graphs in mathematics and computer science. Just like graphs, networks can have weight and flow assigned to its edges, which can be used to represent various interconnected features more accurately. Geometric networks are often used to model road networks and public utility networks, such as electric, gas, and water networks. Network modeling is also commonly employed in transportation planning, hydrology modeling, and infrastructure modeling.

Hydrological Modeling

GIS hydrological models can provide a spatial element that other hydrological models lack, with the analysis of variables such as slope, aspect and watershed or catchment area. Terrain analysis is fundamental to hydrology, since water always flows down a slope. As basic terrain analysis of a digital elevation model (DEM) involves calculation of slope and aspect, DEMs are very useful for hydrological analysis. Slope and aspect can then be used to determine direction of surface runoff, and hence flow accumulation for the formation of streams, rivers and lakes. Areas of divergent flow can also give a clear indication of the boundaries of a catchment. Once a flow direction and accumulation matrix has been created, queries can be performed that show contributing or dispersal areas at a certain point. More detail can be added to the model, such as terrain roughness, vegetation types and soil types, which can influence infiltration and evapotranspiration rates, and hence influencing surface flow. One of the main uses of hydrological modeling is in environmental contamination research.

Cartographic Modeling

Dana Tomlin probably coined the term "cartographic modeling" in his PhD dissertation (1983); he later used it in the title of his book, *Geographic Information Systems and Cartographic Modeling* (1990). Cartographic modeling refers to a process where several thematic layers of the same area are produced, processed, and analyzed. Tomlin used raster layers, but the overlay method can be used more generally. Operations on map layers can be combined into algorithms, and eventually into simulation or optimization models.

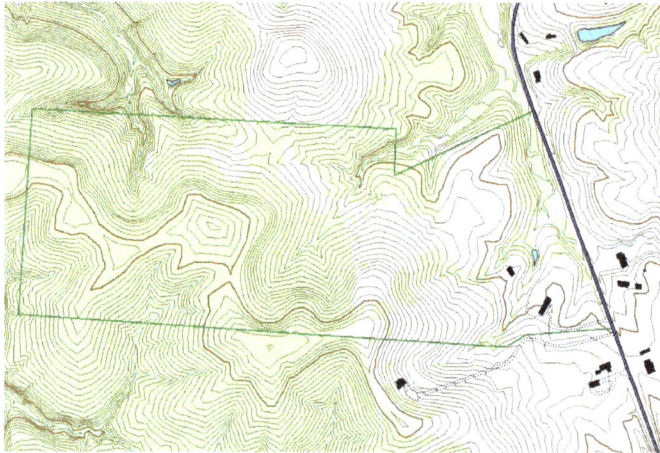

An example of use of layers in a GIS application. In this example, the forest-cover layer (light green) is at the bottom, with the topographic layer over it. Next up is the stream layer, then the boundary layer, then the road layer. The order is very important in order to properly display the final result. Note that the pond layer was located just below the stream layer, so that a stream line can be seen overlying one of the ponds.

Map Overlay

The combination of several spatial datasets (points, lines, or polygons) creates a new output vector dataset, visually similar to stacking several maps of the same region. These overlays are similar to mathematical Venn diagram overlays. A union overlay combines the geographic features and attribute tables of both inputs into a single new output. An intersect overlay defines the area where both inputs overlap and retains a set of attribute fields for each. A symmetric difference overlay defines an output area that includes the total area of both inputs except for the overlapping area.

Data extraction is a GIS process similar to vector overlay, though it can be used in either vector or raster data analysis. Rather than combining the properties and features of both datasets, data extraction involves using a "clip" or "mask" to extract the features of one data set that fall within the spatial extent of another dataset.

In raster data analysis, the overlay of datasets is accomplished through a process known as "local operation on multiple rasters" or "map algebra," through a function that combines the values of each raster's matrix. This function may weigh some inputs more than others through use of an "index model" that reflects the influence of various factors upon a geographic phenomenon.

Geostatistics

Geostatistics is a branch of statistics that deals with field data, spatial data with a continuous index. It provides methods to model spatial correlation, and predict values at arbitrary locations (interpolation).

When phenomena are measured, the observation methods dictate the accuracy of any subsequent analysis. Due to the nature of the data (e.g. traffic patterns in an urban environment; weather patterns over the Pacific Ocean), a constant or dynamic degree of precision is always lost in the measurement. This loss of precision is determined from the scale and distribution of the data collection.

To determine the statistical relevance of the analysis, an average is determined so that points (gradients) outside of any immediate measurement can be included to determine their predicted behavior. This is due to the limitations of the applied statistic and data collection methods, and interpolation is required to predict the behavior of particles, points, and locations that are not directly measurable.

Hillshade model derived from a Digital Elevation Model of the Valestra area in the northern Apennines (Italy)

Interpolation is the process by which a surface is created, usually a raster dataset, through the input of data collected at a number of sample points. There are several forms of interpolation, each which treats the data differently, depending on the properties of the data set. In comparing interpolation methods, the first consideration should be whether or not the source data will change (exact or approximate). Next is whether the method is subjective, a human interpretation, or objective. Then there is the nature of transitions between points: are they abrupt or gradual. Finally, there is whether a method is global (it uses the entire data set to form the model), or local where an algorithm is repeated for a small section of terrain.

Interpolation is a justified measurement because of a spatial autocorrelation principle that recognizes that data collected at any position will have a great similarity to, or influence of those locations within its immediate vicinity.

Digital elevation models, triangulated irregular networks, edge-finding algorithms, Thiessen polygons, Fourier analysis, (weighted) moving averages, inverse distance weighting, kriging, spline, and trend surface analysis are all mathematical methods to produce interpolative data.

Address Geocoding

Geocoding is interpolating spatial locations (X,Y coordinates) from street addresses or any other spatially referenced data such as ZIP Codes, parcel lots and address locations. A reference theme is required to geocode individual addresses, such as a road centerline file with address ranges. The individual address locations have historically been interpolated, or estimated, by examining address ranges along a road segment. These are usually provided in the form of a table or database. The software will then place a dot approximately where that address belongs along the segment of centerline. For example, an address point of 500 will be at the midpoint of a line segment that starts with address 1 and ends with address 1,000. Geocoding can also be applied against actual parcel data, typically from municipal tax maps. In this case, the result of the geocoding will be an actually positioned space as opposed to an interpolated point. This approach is being increasingly used to provide more precise location information.

Reverse Geocoding

Reverse geocoding is the process of returning an estimated street address number as it relates to a given coordinate. For example, a user can click on a road centerline theme (thus providing a coordinate) and have information returned that reflects the estimated house number. This house number is interpolated from a range assigned to that road segment. If the user clicks at the midpoint of a segment that starts with address 1 and ends with 100, the returned value will be somewhere near 50. Note that reverse geocoding does not return actual addresses, only estimates of what should be there based on the predetermined range.

Multi-criteria Decision Analysis

Coupled with GIS, multi-criteria decision analysis methods support decision-makers in analysing a set of alternative spatial solutions, such as the most likely ecological habitat for restoration, against multiple criteria, such as vegetation cover or roads. MCDA uses decision rules to aggregate the criteria, which allows the alternative solutions to be ranked or prioritised. GIS MCDA may reduce costs and time involved in identifying potential restoration sites.

Data Output and Cartography

Cartography is the design and production of maps, or visual representations of spatial data. The vast majority of modern cartography is done with the help of computers, usually using GIS but production of quality cartography is also achieved by importing layers into a design program to refine it. Most GIS software gives the user substantial control over the appearance of the data.

Cartographic work serves two major functions:

First, it produces graphics on the screen or on paper that convey the results of analysis to the people who make decisions about resources. Wall maps and other graphics can be generated, allowing the viewer to visualize and thereby understand the results of analyses or simulations of potential events. Web Map Servers facilitate distribution of generated maps through web browsers using various implementations of web-based application programming interfaces (AJAX, Java, Flash, etc.).

Second, other database information can be generated for further analysis or use. An example would be a list of all addresses within one mile (1.6 km) of a toxic spill.

Graphic Display Techniques

Traditional maps are abstractions of the real world, a sampling of important elements portrayed on a sheet of paper with symbols to represent physical objects. People who use maps must interpret these symbols. Topographic maps show the shape of land surface with contour lines or with shaded relief.

Today, graphic display techniques such as shading based on altitude in a GIS can make relationships among map elements visible, heightening one's ability to extract and analyze information. For example, two types of data were combined in a GIS to produce a perspective view of a portion of San Mateo County, California.

- The digital elevation model, consisting of surface elevations recorded on a 30-meter horizontal grid, shows high elevations as white and low elevation as black.

- The accompanying Landsat Thematic Mapper image shows a false-color infrared image looking down at the same area in 30-meter pixels, or picture elements, for the same coordinate points, pixel by pixel, as the elevation information.

A GIS was used to register and combine the two images to render the three-dimensional perspective view looking down the San Andreas Fault, using the Thematic Mapper image pixels, but shaded using the elevation of the landforms. The GIS display depends on the viewing point of the observer and time of day of the display, to properly render the shadows created by the sun's rays at that latitude, longitude, and time of day.

An archeochrome is a new way of displaying spatial data. It is a thematic on a 3D map that is applied to a specific building or a part of a building. It is suited to the visual display of heat-loss data.

Spatial ETL

Spatial ETL tools provide the data processing functionality of traditional Extract, Transform, Load (ETL) software, but with a primary focus on the ability to manage spatial data. They provide GIS users with the ability to translate data between different standards and proprietary formats, whilst geometrically transforming the data en route. These tools can come in the form of add-ins to existing wider-purpose software such as Microsoft Excel.

GIS Data Mining

GIS or spatial data mining is the application of data mining methods to spatial data. Data mining, which is the partially automated search for hidden patterns in large databases, offers great potential benefits for applied GIS-based decision making. Typical applications include environmental monitoring. A characteristic of such applications is that spatial correlation between data measurements require the use of specialized algorithms for more efficient data analysis.

Applications

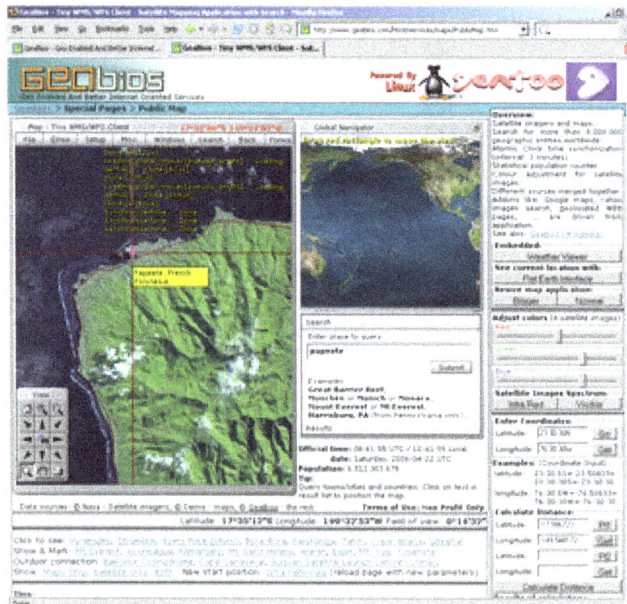

GeaBios – tiny WMS/WFS client (Flash/DHTML)

The implementation of a GIS is often driven by jurisdictional (such as a city), purpose, or application requirements. Generally, a GIS implementation may be custom-designed for an organization. Hence, a GIS deployment developed for an application, jurisdiction, enterprise, or purpose may not be necessarily interoperable or compatible with a GIS that has been developed for some other application, jurisdiction, enterprise, or purpose.

GIS provides, for every kind of location-based organization, a platform to update geographical data without wasting time to visit the field and update a database manually. GIS when integrated with other powerful enterprise solutions like SAP and the Wolfram Language helps creating powerful decision support system at enterprise level.

Many disciplines can benefit from GIS technology. An active GIS market has resulted in lower costs and continual improvements in the hardware and software components of GIS, and usage in the fields of science, government, business, and industry, with applications including real estate, public health, crime mapping, national defense, sustainable development, natural resources, climatology, landscape architecture, archaeology, regional and community planning, transportation and logistics. GIS is also diverging into location-based services, which allows GPS-enabled mobile devices to display their location in relation to fixed objects (nearest restaurant, gas station, fire hydrant) or mobile objects (friends, children, police car), or to relay their position back to a central server for display or other processing.

Open Geospatial Consortium Standards

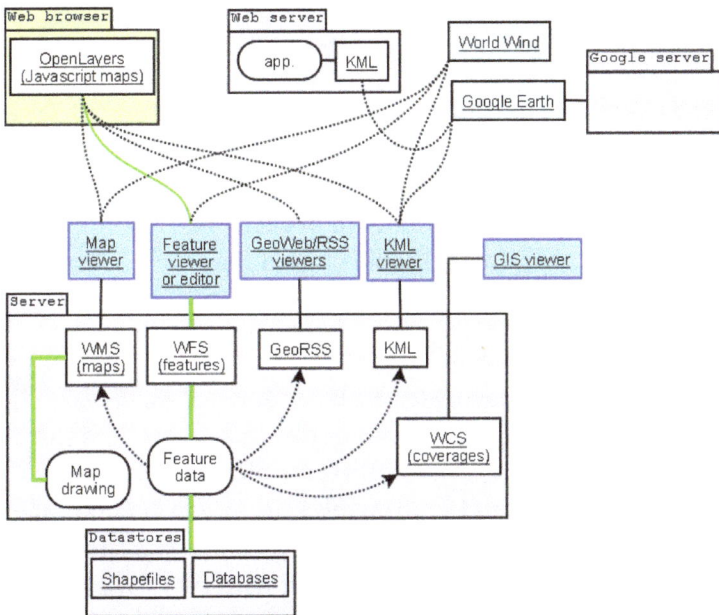

OGC standards help GIS tools communicate.

The Open Geospatial Consortium (OGC) is an international industry consortium of 384 companies, government agencies, universities, and individuals participating in a consensus process to develop publicly available geoprocessing specifications. Open interfaces and protocols defined by OpenGIS Specifications support interoperable solutions that "geo-enable" the Web, wireless and location-based services, and mainstream IT, and empower technology developers to make complex spatial information and services accessible and useful with all kinds of applications. Open Geospatial Consortium protocols include Web Map Service, and Web Feature Service.

GIS products are broken down by the OGC into two categories, based on how completely and accurately the software follows the OGC specifications.

Compliant Products are software products that comply to OGC's OpenGIS Specifications. When a product has been tested and certified as compliant through the OGC Testing Program, the product is automatically registered as "compliant" on this site.

Implementing Products are software products that implement OpenGIS Specifications but have not yet passed a compliance test. Compliance tests are not available for all specifications. Developers can register their products as implementing draft or approved specifications, though OGC reserves the right to review and verify each entry.

Web Mapping

In recent years there has been an proliferation of free-to-use and easily accessible mapping software such as the proprietary web applications Google Maps and Bing Maps, as well as the free and open-source alternative OpenStreetMap. These services give the public access to huge amounts of geographic data.

Some of them, like Google Maps and OpenLayers, expose an API that enable users to create custom applications. These toolkits commonly offer street maps, aerial/satellite imagery, geocoding, searches, and routing functionality. Web mapping has also uncovered the potential of crowdsourcing geodata in projects like OpenStreetMap, which is a collaborative project to create a free editable map of the world.

Adding the Dimension of Time

The condition of the Earth's surface, atmosphere, and subsurface can be examined by feeding satellite data into a GIS. GIS technology gives researchers the ability to examine the variations in Earth processes over days, months, and years. As an example, the changes in vegetation vigor through a growing season can be animated to determine when drought was most extensive in a particular region. The resulting graphic represents a rough measure of plant health. Working with two variables over time would then allow researchers to detect regional differences in the lag between a decline in rainfall and its effect on vegetation.

GIS technology and the availability of digital data on regional and global scales enable such analyses. The satellite sensor output used to generate a vegetation graphic is produced for example by the Advanced Very High Resolution Radiometer (AVHRR). This sensor system detects the amounts of energy reflected from the Earth's surface across various bands of the spectrum for surface areas of about 1 square kilometer. The satellite sensor produces images of a particular location on the Earth twice a day. AVHRR and more recently the Moderate-Resolution Imaging Spectroradiometer (MODIS) are only two of many sensor systems used for Earth surface analysis. More sensors will follow, generating ever greater amounts of data.

In addition to the integration of time in environmental studies, GIS is also being explored for its ability to track and model the progress of humans throughout their daily routines. A concrete example of progress in this area is the recent release of time-specific population data by the U.S. Census. In this data set, the populations of cities are shown for daytime and evening hours highlighting the pattern of concentration and dispersion generated by North American commuting patterns. The manipulation and generation of data required to produce this data would not have been possible without GIS.

Using models to project the data held by a GIS forward in time have enabled planners to test policy decisions using spatial decision support systems.

Semantics

Tools and technologies emerging from the W3C's Data Activity are proving useful for data integration problems in information systems. Correspondingly, such technologies have been proposed as a means to facilitate interoperability and data reuse among GIS applications. and also to enable new analysis mechanisms.

Ontologies are a key component of this semantic approach as they allow a formal, machine-readable specification of the concepts and relationships in a given domain. This in turn allows a GIS to focus on the intended meaning of data rather than its syntax or structure. For example, reasoning that a land cover type classified as *deciduous needleleaf trees* in one dataset is a specialization or subset of land cover type *forest* in another more roughly classified dataset can help a GIS automatically merge the two datasets under the more general land cover classification. Tentative ontologies have been developed in areas related to GIS applications, for example the hydrology ontology developed by the Ordnance Survey in the United Kingdom and the SWEET ontologies developed by NASA's Jet Propulsion Laboratory. Also, simpler ontologies and semantic metadata standards are being proposed by the W3C Geo Incubator Group to represent geospatial data on the web. GeoSPARQL is a standard developed by the Ordnance Survey, United States Geological Survey, Natural Resources Canada, Australia's Commonwealth Scientific and Industrial Research Organisation and others to support ontology creation and reasoning using well-understood OGC literals (GML, WKT), topological relationships (Simple Features, RCC8, DE-9IM), RDF and the SPARQL database query protocols.

Recent research results in this area can be seen in the International Conference on Geospatial Semantics and the Terra Cognita – Directions to the Geospatial Semantic Web workshop at the International Semantic Web Conference.

Implications of GIS in Society

With the popularization of GIS in decision making, scholars have begun to scrutinize the social and political implications of GIS. GIS can also be misused to distort reality for individual and political gain. It has been argued that the production, distribution, utilization, and representation of geographic information are largely related with the social context and has the potential to increase citizen trust in government. Other related topics include discussion on copyright, privacy, and censorship. A more optimistic social approach to GIS adoption is to use it as a tool for public participation.

History of GIS

Mapmaking (representation of geographical information) has evidences to show independent evolution of maps in different parts of the earth. The direct evidence of mapping comes from Middle East in the form of Babylonian Clay Tablets as early as 1000 B.C which depicted earth as a flat circular disk.

Around 200 B.C, Eratosthenes calculated the circumference of earth accurately. Later came, Ptolemy and Al-Idrisi who made remarkable contributions in the field of cartography. Following them were Mercator and Newton, their work paved way for the upcoming cartographers and geographers to better understand the earth and the geographical phenomenon.

Putting layers of data on series of base maps to analyze things geographically has been into existence much longer than the introduction of computers to the geographical world.

The French cartographer Louis-Alexandre Berthier had drawn the maps of the Battle of Yorktown (1781) that contained hinged overlays to show troop movements.

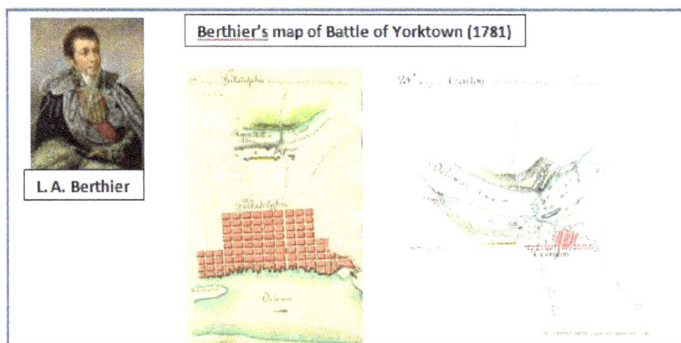

L. A. Berthier

Berthier's map of Battle of Yorktown (1781)

Superimposition of topography, geology, population and traffic flow on the same base map has been shown in the Atlas to Accompany the Second report of the Irish Railway Commissioners.

Dr. John Snow showed the locations of death by cholera on a map to track the source of outbreak of cholera in Central London in September, 1854.

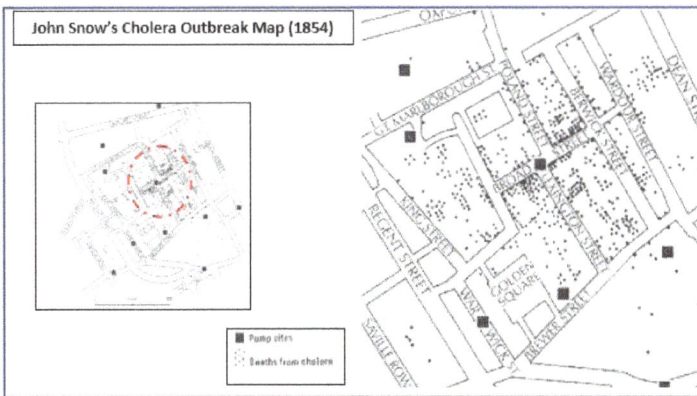

The introduction of computers in the field of geography was a positive step towards understanding and learning the subject better. Change in cartographic analysis due to improved graphics, development of theories of spatial processes in economic and social geography, anthropology and regional science, increased social awareness and improvement in education. The integrated transportation plans of Detroit, Chicago during the period of 1950s and 1960s used information on routes, origin, destination, and time to produce the maps of traffic flow and volume is an example of integration of computer technology with geographical data.

Application of GIS

GIS is involved in various areas. These include topographical mapping, socioeconomic and environment modeling, and education. The role of GIS is best illustrated with respect to some of the representative application areas that are mentioned below:

Tax Mapping: Raising revenue from property taxes is one of the important functions of the government agencies. The amount of tax payable depends on the value of the land and the property. The correct assessment of value of land and property determines the equitable distribution of the community tax. A tax assessor has to evaluate new properties and respond to the existing property valuation. To evaluate taxes the assessor uses details on current market rents, sale, maintenance, insurance and other expenses. Managing as well as analyzing all this information simultaneously is time consuming and hence comes the need of GIS. Information about property with its geographical location and boundary is managed by GIS. Land units stored in parcel database can be linked to their properties. Querying the GIS database can locate similar type of properties in an area. The characteristics of these properties can then be compared and valuation can be easily done.

Business: Approximately 80 percent of all business data are related to location. Businesses manage a world of information about sales, customers, inventory, demographic profiles etc. Demographic analysis is the basis for many other business functions: customer service, site analysis, and marketing. Understanding your customers and their socioeconomic and purchasing behavior is essential for making good business decisions. A GIS with relevant data such as number of consumers, brands and sites they go for shopping can give any business unit a fair idea whether their unit if set up is going to work at a particular location the way they want it to run.

Logistics: Logistics is a field that takes care of transporting goods from one place to another and finally delivering them to their destinations. It is necessary for the shipping companies to know where their warehouses should be located, which routes should the transport follow that ensures minimum time and expenditures to deliver the parcels to their destinations. All such logistics decisions need GIS support.

Emergency evacuation: The occurrence of disasters is unpredictable. We as humans are unable to tell when, where and what magnitude of disaster is going to emerge and therefore solely depend on disaster preparedness as safety measures. It is important to know in which area the risk is higher, the number of individuals inhabiting that place, the routes by which the vehicles would move to help in evacuating the individuals. Thus preparing an evacuation plan needs GIS implementation.

Environment: GIS is being increasingly involved in mapping the habitat loss, urban sprawl, land-use change etc. Mapping such phenomena need historical landuse data, anthropogenic effects which greatly affect these phenomena are also brought into GIS domain. GIS models are then run to make predictions for the future.

Geostatistics

Geostatistics is a branch of statistics focusing on spatial or spatiotemporal datasets. Developed originally to predict probability distributions of ore grades for mining operations, it is currently applied in diverse disciplines including petroleum geology, hydrogeology, hydrology, meteorology, oceanography, geochemistry, geometallurgy, geography, forestry, environmental control, landscape ecology, soil science, and agriculture (esp. in precision farming). Geostatistics is applied in varied branches of geography, particularly those involving the spread of diseases (epidemiology), the practice of commerce and military planning (logistics), and the development of efficient spatial networks. Geostatistical algorithms are incorporated in many places, including geographic information systems (GIS) and the R statistical environment.

Background

Geostatistics is intimately related to interpolation methods, but extends far beyond simple interpolation problems. Geostatistical techniques rely on statistical models that are based on random function (or random variable) theory to model the uncertainty associated with spatial estimation and simulation.

A number of simpler interpolation methods/algorithms, such as inverse distance weighting, bilinear interpolation and nearest-neighbor interpolation, were already well known before geostatistics. Geostatistics goes beyond the interpolation problem by considering the studied phenomenon at unknown locations as a set of correlated random variables.

Let $Z(x)$ be the value of the variable of interest at a certain location x. This value is unknown (e.g. temperature, rainfall, piezometric level, geological facies, etc.). Although there exists a value at location x that could be measured, geostatistics considers this value as random since it was not measured, or has not been measured yet. However, the randomness of $Z(x)$ is not complete, but defined by a cumulative distribution function (CDF) that depends on certain information that is known about the value $Z(x)$:

$$F(z, \mathbf{x}) = \text{Prob}\{Z(\mathbf{x}) \leqslant z \,|\, \text{information}\}.$$

Typically, if the value of Z is known at locations close to x (or in the neighborhood of x) one can constrain the CDF of $Z(x)$ by this neighborhood: if a high spatial continuity is assumed, $Z(x)$ can only have values similar to the ones found in the neighborhood. Conversely, in the absence of spatial continuity $Z(x)$ can take any value. The spatial continuity of the random variables is described by a model of spatial continuity that can be either a parametric function in the case of variogram-based geostatistics, or have a non-parametric form when using other methods such as multiple-point simulation or pseudo-genetic techniques.

By applying a single spatial model on an entire domain, one makes the assumption that Z is a stationary process. It means that the same statistical properties are applicable on the entire domain. Several geostatistical methods provide ways of relaxing this stationarity assumption.

In this framework, one can distinguish two modeling goals:

1. Estimating the value for $Z(x)$, typically by the expectation, the median or the mode of the CDF $f(z,x)$. This is usually denoted as an estimation problem.

2. Sampling from the entire probability density function $f(z,x)$ by actually considering each possible outcome of it at each location. This is generally done by creating several alternative maps of Z, called realizations. Consider a domain discretized in N grid nodes (or pixels). Each realization is a sample of the complete N-dimensional joint distribution function

$$F(\mathbf{z}, \mathbf{x}) = \text{Prob}\{Z(\mathbf{x}_1) \leqslant z_1, Z(\mathbf{x}_2) \leqslant z_2, ..., Z(\mathbf{x}_N) \leqslant z_N\}.$$

In this approach, the presence of multiple solutions to the interpolation problem is acknowledged. Each realization is considered as a possible scenario of what the real variable could be. All associated workflows are then considering ensemble of realizations, and consequently ensemble of predictions that allow for probabilistic forecasting. Therefore, geostatistics is often used to generate or update spatial models when solving inverse problems.

A number of methods exist for both geostatistical estimation and multiple realizations approaches. Several reference books provide a comprehensive overview of the discipline.

References

- Goodchild, Michael F (2010). "Twenty years of progress: GIScience in 2010". Journal of Spatial Information Science (1). doi:10.5311/JOSIS.2010.1.2

- Lovison-Golob, Lucia. "Howard T. Fisher". Harvard University. Archived from the original on 2007-12-13. Retrieved 2007-06-09

- Haque, Akhlaque (2015). Surveillance, Transparency and Democracy: Public Administration in the Information Age. Tuscaloosa, AL: University of Alabama Press. pp. 70–73. ISBN 978-0817318772

- "Rapport sur la marche et les effets du choléra dans Paris et le département de la Seine. Année 1832". Gallica. Retrieved 10 May 2012

- Ma, Y.; Guo, Y.; Tian, X.; Ghanem, M. (2011). "Distributed Clustering-Based Aggregation Algorithm for Spatial Correlated Sensor Networks". IEEE Sensors Journal. 11 (3): 641. doi:10.1109/JSEN.2010.2056916

- Tomlin, C. Dana (1990). Geographic information systems and cartographic modeling. Prentice Hall series in geographic information science. Prentice Hall. Retrieved 2017-01-05

- Sinton, Diana Stuart; Lund, Jennifer J., eds. (2007). Understanding place: GIS and mapping across the curriculum. Redlands, CA: ESRI Press. ISBN 9781589481497. OCLC 70866933

- The Remarkable History of GIS - Geographical Information Systems."The Remarkable History of GIS". Retrieved 2015-05-05

Maps: Design and Scaling

Maps are a visual representation of objects and regions that are found on Earth. Maps can be divided into scale based maps and purpose scale maps. Map scaling has improved with computer aided designs and graphics. Map design and scaling can best be understood in confluence with the major topics listed in the following section.

Maps and Map Scales

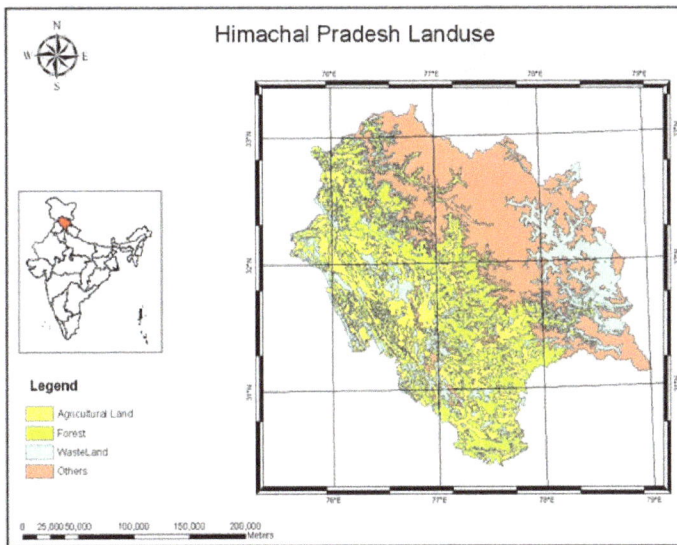

Map showing Landuse of Himachal Pradesh

A map is a two dimensional representation of earth surface which uses graphics to convey geographical information. It describes the geographical location of features and the relationship between them. Maps are fundamental to society. Cartography refers to the art and science of map preparation. Though, the earliest of the maps were technically not as impressive as they are today but they certainly highlighted their role in communicating information about the location and spatial characteristics of the natural world and of society and culture. The new discoveries in Science and Geography fortified maps with facts and technical details. Improvements in the fields of Geodesy, Surveying and Cartography helped in bringing the maps to their present form. The digital technology has altered the way of creating, presenting and distributing the geographic information. The conventional cartography is now getting replaced by computer aided designs and graphics, and the analog maps (paper maps) by digital maps. The growing field of technology promises to bring more advances to Cartography to render maps and allied services that serve the society in a better manner.

History of Maps

The oldest known maps are from Middle East in the form of Babylonian Clay Tablet dating as early as 1000 B.C and depicted earth as a flat circular disk.

Babylonian clay tablet

Interpretive redrawing of the clay tablet

The clay tablet shows a flat, round world with Babylonia in the centre. The other regions of the world are beyond the ocean encircling Babylonia.

Around 150 AD, Ptolemy depicted the old world from 60° N to 30° S latitude on a map. The first use of longitudinal and latitudinal lines on a map along with the specification of terrestrial sites through observations of the celestial sphere is regarded as one of the important contribution of Ptolemy in the field of cartography.

Ptolemy's world map

Muhammad al-Idrisi is one of the cartographers of the medieval period who created the world map by combining the information on Africa, Indian Ocean and the Far East provided by merchants and geographers. His map was considered the most accurate world map for the next three centuries.

Al Idrisi's world map

After the voyages by Columbus and others to the new world, the full world map started to appear in the early 16th Century. Martin Waldseemüller in 1507 is credited with the creation of the first true world map. The map used Ptolemaic projection and was the first one to use the name America for the New World.

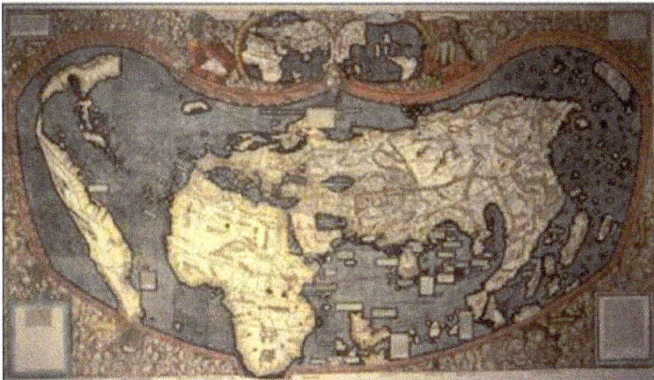

Waldseemuller world map

Gerardus Mercator, a cartographer of the mid-16th century developed a cylindrical projection that is still widely used for navigation charts and global maps. He then published a map of the world in 1569 based on this projection.

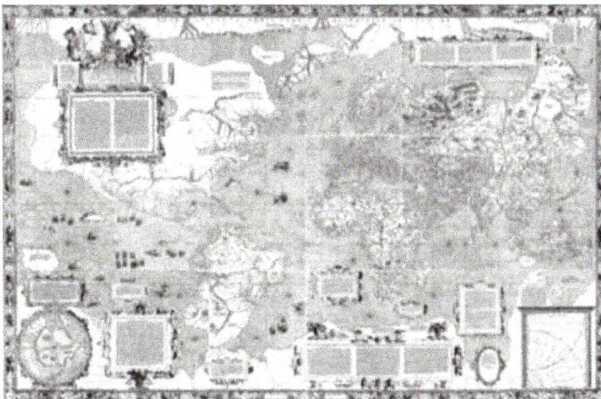

Mercator's world map

Maps during the 17th, 18th and 19th centuries became increasingly accurate and factual with the application of scientific methods. With the advent of GIS during mid 1960s, understanding and representation of geographical phenomenon improved significantly. Maps now have improved graphics, and spatial relation visualization.

Types of Maps

The maps can be classified on the following criteria:

- Scale

- Purpose

Scale is important for correct representation of geographical features and phenomenon. Different features require different scales for their display. For example preparation of a cadastral map of a village and the soil map of a state would use different scale for representing the information. According to scale, maps can be classified as follows:

a. Cadastral : These maps register the ownership of land property. They are prepared by government to realize tax and revenue. A village map is an example of cadastral map which is drawn on a scale of 16 inches to the mile or 32 inches to the mile.

b. Topographical: Topographical maps are prepared on fairly large scale and are based on precise survey. They don't reveal land parcels but show topographic forms such as relief, drainage, forest, village, towns etc. The scale of these maps varied conventionally from 1/4 inch to the mile to one inch to the mile. The topographical maps of different countries have varying scales.

- Topographical survey map of British Ordnance Survey are one inch maps.

- The scale of European toposheets varies from 1:25000 to 1:100000.

- USA toposheets are drawn on the scale 1:62500 and 1:125000.

- The international map which is a uniform map of the world is produced on the scale of 1:1000000.

c. Chorographical/Atlas: Drawn on a very small scale, atlas maps give a generalized view of physical, climatic and economic conditions of different regions of the world. The scale of atlas map is generally greater than 1:1000,000.

On the basis of Purpose or the content, the maps can be classified as follows:

a. Natural Maps:

These maps represent natural features and the processes associated with them. Given below is the list of some such maps:

Astronomical map: It refers to the cartographic representation of the heavenly bodies such as galaxies, stars, planets, moon etc.

Geological map: A map that represents the distribution of different type of rocks and surficial deposits on the Earth.

Relief map: A map that depicts the terrain and indicates the bulges and the depressions present on the surface.

Climate map: A climate map is a depiction of prevailing weather patterns in a given area. These maps can show daily weather conditions, average monthly or seasonal weather conditions of an area.

Vegetation map: It shows the natural flora of an area.

Soil map: A soil map describes the soil cover present in an area.

b. Cultural Maps:

These maps tell about the cultural patterns designed over the surface of the earth. They describe the activities of man and related processes. Given below is the list of such maps:

Political map: A map that shows the boundaries of states, boundaries between different political units of the world or of a particular country which mark the areas of respective political jurisdiction

Military map: A military map contains information about routes, points, security and battle plans.

Historical map: A map having historical events symbolized on it.

Social map: A map giving information about the tribes, languages and religions of an area.

Land-utilization map: A map describing the land and the ongoing activities on it.

Communication map: A map showing means of communication such as railways, road, airways etc.

Population map: A map showing distribution of human beings over an area.

Cadastre

A cadastre (also spelled cadaster), using a cadastral survey or cadastral map, is a comprehensive register of the real estate or real property's metes-and-bounds of a country.

Cadastral map of the village of Pielnia, 1852, Austrian Empire

In most countries, legal systems have developed around the original administrative systems and use the cadastre to define the dimensions and location of land parcels described in legal documentation. The cadastre is a fundamental source of data in disputes and lawsuits between landowners.

In the United States, Cadastral Survey within the Bureau of Land Management (BLM) maintains records of all public lands. Such surveys often require detailed investigation of the history of land use, legal accounts, and other documents.

Definition

A cadastre commonly includes details of the ownership, the tenure, the precise location (therefore GNSS coordinates are not used due to errors such as multipath), the dimensions (and area), the cultivations if rural, and the value of individual parcels of land. Cadastres are used by many nations around the world, some in conjunction with other records, such as a title register.

The International Federation of Surveyors defines cadastre as follows:

A Cadastre is normally a parcel based, and up-to-date land information system containing a record of interests in land (e.g. rights, restrictions and responsibilities). It usually includes a geometric description of land parcels linked to other records describing the nature of the interests, the ownership or control of those interests, and often the value of the parcel and its improvements.

Etymology

The word *cadastre* came into English through French from Late Latin *capitastrum*, a register of the poll tax, and the Greek, a list or register, which literally means, "down the line", in the sense of "line by line" along the directions and distances between the corners mentioned and marked by monuments in the metes and bounds.

The word forms the adjective *cadastral*, used in public administration, primarily for ownership and taxation purposes. The terminology for cadastral divisions may include counties, parishes, ridings, hundreds, sections, lots, blocks and city blocks.

History

Some of the earliest cadasters were ordered by Roman Emperors to recover state owned lands that had been appropriated by private individuals, and thereby recover income from such holdings. One such cadaster was done in 77 AD in Campania, a surviving stone marker of the survey reads "The Emperor Vespasian, in the eighth year of his tribunician power, so as to restore the state lands which the Emperor Augustus had given to the soldiers of Legion II Gallica, but which for some years had been occupied by private individuals, ordered a survey map to be set up with a record on each 'century' of the annual rental". In this way Vespasian was able to reimpose taxation formerly uncollected on these lands.

With the fall of Rome the use of cadastral maps effectively discontinued. Medieval practice used written descriptions of the extent of land rather than using more precise surveys. Only in the sixteenth and early seventeenth centuries did the use of cadastral maps resume, beginning in the Netherlands. With the emergence of capitalism in Renaissance Europe the need for cadastral maps reemerged as a tool to determine and express control of land as a means of production. This took place first privately in land disputes and later spread to governmental practice as a means of more precise tax assessment.

Cadastral Surveys

BLM cadastral survey marker from 1992 in San Xavier, Arizona

Cadastral surveys document the boundaries of land ownership, by the production of documents, diagrams, sketches, plans (plats in USA), charts, and maps. They were originally used to ensure reliable facts for land valuation and taxation. An example from early England is the Domesday Book in 1086. Napoleon established a comprehensive cadastral system for France that is regarded as the forerunner of most modern versions.

The Public Lands Survey System is a cadastral survey of the United States originating in legislation from 1785, after international recognition of the United States. The Dominion Land Survey is a similar cadastral survey conducted in Western Canada begun in 1871 after the creation of the Dominion of Canada in 1867. Both cadastral surveys are made relative to principal meridian and baselines. These cadastral surveys divided the surveyed areas into townships, square land areas of approximately 36 square miles (six miles by six miles; some very early surveys in Ohio created 25 square mile townships when the design of the system was being explored). These townships are divided into sections, each approximately one-mile square. Unlike in Europe this cadastral survey largely preceded settlement and as a result greatly influenced settlement patterns. Properties are generally rectangular, boundary lines often run on cardinal bearings, and parcel dimensions are often in fractions or multiples of chains. Land descriptions in Western North America are principally based on these land surveys.

Cadastral survey information is often a base element in Geographic Information Systems (GIS) or Land Information Systems (LIS) used to assess and manage land and built infrastructure. Such systems are also employed on a variety of other tasks, for example, to track long-term changes over time for geological or ecological studies, where land tenure is a significant part of the scenario.

Cadastral Map

A *cadastral map* is a map that shows the boundaries and ownership of land parcels. Some cadastral maps show additional details, such as survey district names, unique identifying numbers for parcels, certificate of title numbers, positions of existing structures, section or lot numbers and their respective areas, adjoining and adjacent street names, selected boundary dimensions and references to prior maps.

The U.S. NOAA Coastal Services Center (CSC) has released a cadastre web tool to illustrate offshore wind power suitability of Eastern seaboard areas.

Cadastral Documentation

Comprises the documentary materials submitted to cadastre or land administration offices for renewal of cadastral recordings. Cadastral documentation is kept in paper and/or electronic form. Jurisdiction statutes and further provisions specify the content and form of the documentation, as well as the person(s) authorized to prepare and sign the documentation, including concerned parties (owner, etc.), licensed surveyors and legal advisors. The office concerned reviews the submitted information; if the documentation does not comply with stated provisions, the office may set a deadline for the applicant to submit complete documentation.

Despite the age of Cadastre, the notion of Cadastral documentation emerged late in the English language, as the institution of cadaster developed outside English-speaking countries. In a Danish textbook, one out of 15 s regards the Form and content of documents concerning subdivision, etc. Early textbooks of international scope focused on recordings in terms of land registration and technical aspects of cadastral survey, yet note that 'cadastral surveying has been carried out within a tight framework of legislation'. With the view of assessing transaction costs, a European project: Modelling real property transactions (2001-2005) charted procedures for the transfer of ownership and other rights in land and buildings. Cadastral documentation is described, e.g. for Finland as follows '8. Surveyor draws up cadstral map and cadastral documents .. 10. Surveyor sends cadastral documents to cadastral authority.' In Australia, similar activities are referred to as 'lodgement of plans of subdivision at land titles offices'

Topographic Map

In modern mapping, a topographic map is a type of map characterized by large-scale detail and quantitative representation of relief, usually using contour lines, but historically using a variety of methods. Traditional definitions require a topographic map to show both natural and man-made features. A topographic map is typically published as a map series, made up of two or more map sheets that combine to form the whole map. A contour line is a line connecting places of equal elevation.

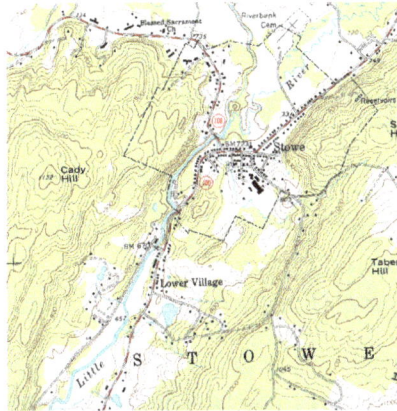

A topographic map with contour lines

Part of the same map in a perspective shaded relief view illustrating how the contour lines follow the

terrain

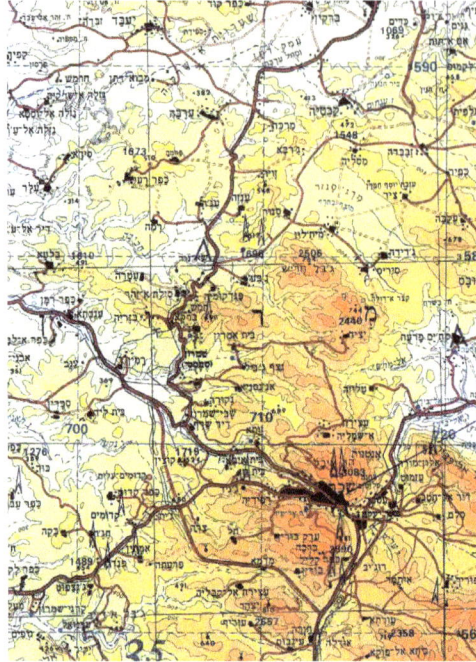

Section of topographical map of Nablus area (West Bank) with contour lines at 100-meter intervals.
Heights are colour-coded

Natural Resources Canada provides this description of topographic maps:

These maps depict in detail ground relief (landforms and terrain), drainage (lakes and rivers), forest cover, administrative areas, populated areas, transportation routes and facilities (including roads and railways), and other man-made features.

Other authors define topographic maps by contrasting them with another type of map; they are distinguished from smaller-scale "chorographic maps" that cover large regions, "planimetric maps" that do not show elevations, and "thematic maps" that focus on specific topics.

However, in the vernacular and day to day world, the representation of relief (contours) is popularly held to define the genre, such that even small-scale maps showing relief are commonly (and erroneously, in the technical sense) called "topographic".

The study or discipline of topography is a much broader field of study, which takes into account all natural and man-made features of terrain.

History

Topographic maps are based on topographical surveys. Performed at large scales, these surveys are called topographical in the old sense of topography, showing a variety of elevations and landforms. This is in contrast to older cadastral surveys, which primarily show property and governmental boundaries. The first multi-sheet topographic map series of an entire country, the *Carte géométrique de la France*, was completed in 1789. The Great Trigonometric Survey of India, started by the East India Company in 1802, then taken over by the British Raj after 1857 was notable as a successful effort on a larger scale and for accurately determining heights of Himalayan peaks from viewpoints over one hundred miles distant.

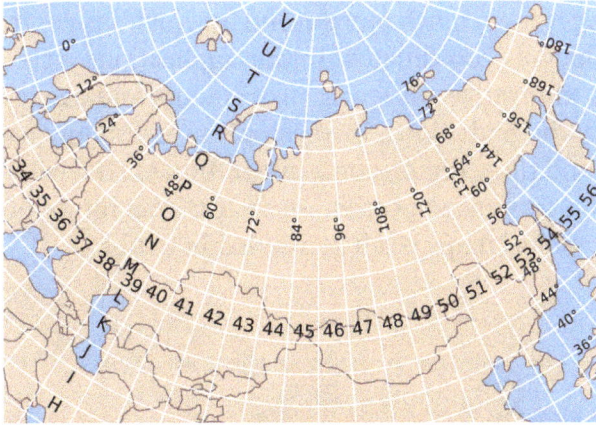

Global indexing system first developed for *International Map of the World*

Topographic surveys were prepared by the military to assist in planning for battle and for defensive emplacements (thus the name and history of the United Kingdom's Ordnance Survey). As such, elevation information was of vital importance.

As they evolved, topographic map series became a national resource in modern nations in planning infrastructure and resource exploitation. In the United States, the national map-making function which had been shared by both the Army Corps of Engineers and the Department of the Interior migrated to the newly created United States Geological Survey in 1879, where it has remained since.

1913 saw the beginning of the International Map of the World initiative, which set out to map all of Earth's significant land areas at a scale of 1:1 million, on about one thousand sheets, each covering four degrees latitude by six or more degrees longitude. Excluding borders, each sheet was 44 cm high and (depending on latitude) up to 66 cm wide. Although the project eventually foundered, it left an indexing system that remains in use.

By the 1980s, centralized printing of standardized topographic maps began to be superseded by databases of coordinates that could be used on computers by moderately skilled end users to view or print maps with arbitrary contents, coverage and scale. For example, the Federal government of the United States' *TIGER* initiative compiled interlinked databases of federal, state and local political borders and census enumeration areas, and of roadways, railroads, and water features with support for locating street addresses within street segments. TIGER was developed in the 1980s and used in the 1990 and subsequent decennial censuses. Digital elevation models (*DEM*) were also compiled, initially from topographic maps and stereographic interpretation of aerial photographs and then from satellite photography and radar data. Since all these were government projects funded with taxes and not classified for national security reasons, the datasets were in the public domain and freely usable without fees or licensing.

TIGER and DEM datasets greatly facilitated Geographic information systems and made the Global Positioning System much more useful by providing context around locations given by the technology as coordinates. Initial applications were mostly professionalized forms such as innovative surveying instruments and agency-level GIS systems tended by experts. By the mid-1990s, increasingly user-friendly resources such as online mapping in two and three dimensions, integration of GPS with mobile phones and automotive navigation systems appeared. As of 2011, the future of standardized, centrally printed topographical maps is left somewhat in doubt.

Uses

Curvimeter used to measure the length of a curve

Topographic maps have multiple uses in the present day: any type of geographic planning or large-scale architecture; earth sciences and many other geographic disciplines; mining and other earth-based endeavours; civil engineering and recreational uses such as hiking and orienteering.

Conventions

The various features shown on the map are represented by conventional signs or symbols. For example, colors can be used to indicate a classification of roads. These signs are usually explained in the margin of the map, or on a separately published characteristic sheet.

Topographic maps are also commonly called *contour maps* or *topo maps*. In the United States, where the primary national series is organized by a strict 7.5-minute grid, they are often called *topo quads* or quadrangles.

Topographic maps conventionally show topography, or land contours, by means of contour lines. Contour lines are curves that connect contiguous points of the same altitude (isohypse). In other words, every point on the marked line of 100 m elevation is 100 m above mean sea level.

These maps usually show not only the contours, but also any significant streams or other bodies of water, forest cover, built-up areas or individual buildings (depending on scale), and other features and points of interest.

Today, topographic maps are prepared using photogrammetric interpretation of aerial photography, lidar and other Remote sensing techniques. Older topographic maps were prepared using traditional surveying instruments.

Geologic Map

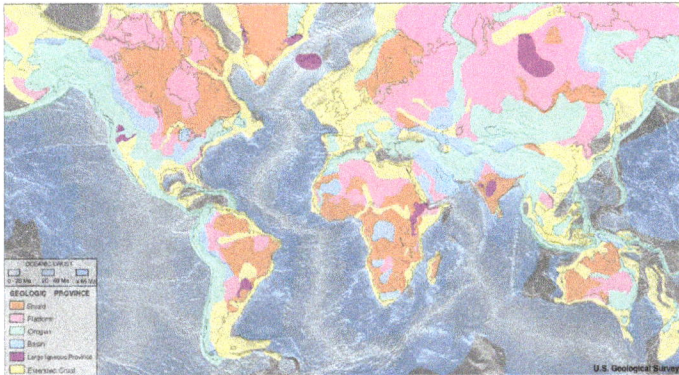

Mapped global geologic provinces

A geologic map or geological map is a special-purpose map made to show geological features. Rock units or geologic strata are shown by color or symbols to indicate where they are exposed at the surface. Bedding planes and structural features such as faults, folds, foliations, and lineations are shown with strike and dip or trend and plunge symbols which give these features' three-dimensional orientations.

Stratigraphic contour lines may be used to illustrate the surface of a selected stratum illustrating the subsurface topographic trends of the strata. Isopach maps detail the variations in thickness of stratigraphic units. It is not always possible to properly show this when the strata are extremely fractured, mixed, in some discontinuities, or where they are otherwise disturbed.

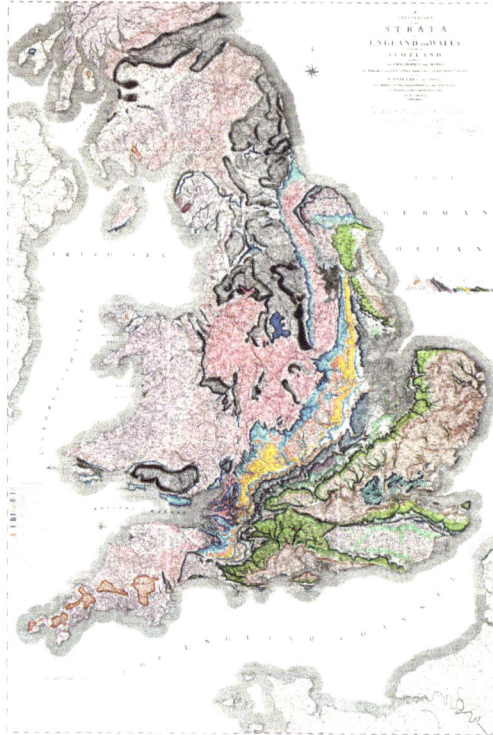

William Smith's geologic map

Symbols

Lithologies

Rock units are typically represented by colors. Instead of (or in addition to) colors, certain symbols can be used. Different geologic mapping agencies and authorities have different standards for the colors and symbols to be used for rocks of differing types and ages.

Orientations

Geologists take two major types of orientation measurements (using a hand compass like a Brunton compass): orientations of planes and orientations of lines. Orientations of planes are often measured as a "strike" and "dip", while orientations of lines are often measured as a "trend" and "plunge".

Strike and dip symbols consist of a long "strike" line, which is perpendicular to the direction of greatest slope along the surface of the bed, and a shorter "dip" line on side of the strike line where the bed is going downwards. The angle that the bed makes with the horizontal, along the dip direction, is written next to the dip line. In the azimuthal system, strike and dip are often given as "strike/dip" (for example: 270/15, for a strike of west and a dip of 15 degrees below the horizontal).

A standard Brunton Geological compass, used commonly by geologists

Trend and plunge are used for linear features, and their symbol is a single arrow on the map. The arrow is oriented in the downgoing direction of the linear feature (the "trend") and at the end of the arrow, the number of degrees that the feature lies below the horizontal (the "plunge") is noted. Trend and plunge are often notated as PLUNGE $\rightarrow$ TREND (for example: 34 $\rightarrow$ 86 indicates a feature that is angled at 34 degrees below the horizontal at an angle that is just East of true South).

History

The oldest preserved geologic map is the Turin papyrus (1150 BCE), which shows the location of building stone and gold deposits in Egypt.

The earliest geologic map of the modern era is the 1771 "Map of Part of Auvergne, or figures of, The Current of Lava in which Prisms, Balls, Etc. are Made from Basalt. To be used with Mr. Demarest's theories of this hard basalt. Engraved by Messr. Pasumot and Daily, Geological Engineers of the King." This map is based on Nicolas Desmarest's 1768 detailed study of the geology and eruptive history of the Auvergne volcanoes and a comparison with the columns of the Giant's Causeway of Ireland. He identified both landmarks as features of extinct volcanoes. The 1798 report was incorporated in the 1771 (French) Royal Academy of Science compendium.

The first geological map of the U.S. was produced in 1809 by William Maclure. In 1807, Maclure undertook the self-imposed task of making a geological survey of the United States. He traversed and mapped nearly every state in the Union. During the rigorous two-year period of his survey, he crossed and recrossed the Allegheny Mountains some 50 times. Maclure's map shows the distribution of five classes of rock in what are now only the eastern states of the present-day US.

The first geologic map of Great Britain was created by William Smith in 1815.

Maps and Mapping Around the Globe

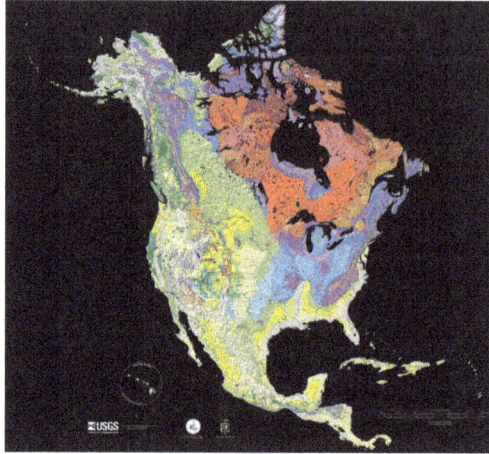

Geologic map of North America superimposed on a shaded relief map

United States

In the United States, geologic maps are usually superimposed over a topographic map (and at times over other base maps) with the addition of a color mask with letter symbols to represent the kind of geologic unit. The color mask denotes the exposure of the immediate bedrock, even if obscured by soil or other cover. Each area of color denotes a geologic unit or particular rock formation (as more information is gathered new geologic units may be defined). However, in areas where the bedrock is overlain by a significantly thick unconsolidated burden of till, terrace sediments, loess deposits, or other important feature, these are shown instead. Stratigraphic contour lines, fault lines, strike and dip symbols, are represented with various symbols as indicated by the map key. Whereas topographic maps are produced by the United States Geological Survey in conjunction with the states, geologic maps are usually produced by the individual states. There are almost no geologic map resources for some states, while a few states, such as Kentucky and Georgia, are extensively mapped geologically. Technically A map that uses colors.

United Kingdom

In the United Kingdom the term *geological map* is used. The UK and Isle of Man have been extensively mapped by the British Geological Survey (BGS) since 1835; a separate Geological Survey of Northern Ireland (drawing on BGS staff) has operated since 1947.

Two 1:625,000 scale maps cover the basic geology for the UK. More detailed sheets are available at scales of 1:250,000, 1:50,000 and 1:10,000. The 1:625,000 and 1:250,000 scales show both onshore and offshore geology (the 1:250,000 series covers the entire UK continental shelf), whilst other scales generally cover exposures on land only.

Sheets of all scales (though not for all areas) fall into two categories:

1. Superficial deposit maps (previously known as *solid and drift* maps) show both bedrock *and* the deposits on top of it.

2. Bedrock maps (previously known as *solid* maps) show the underlying rock, without superficial deposits.

The maps are superimposed over a topographic map base produced by Ordnance Survey (OS), and use symbols to represent fault lines, strike and dip or geological units, boreholes etc. Colors are used to represent different geological units. Explanatory booklets (memoirs) are produced for many sheets at the 1:50,000 scale.

Small scale thematic maps (1:1,000,000 to 1:100,000) are also produced covering geochemistry, gravity anomaly, magnetic anomaly, groundwater, etc.

Although BGS maps show the British national grid reference system and employ an OS base map, sheet boundaries are not based on the grid. The 1:50,000 sheets originate from earlier 'one inch to the mile' (1:63,360) coverage utilising the pre-grid Ordnance Survey One Inch Third Edition as the base map. Current sheets are a mixture of modern field mapping at 1:10,000 redrawn at the 1:50,000 scale and older 1:63,360 maps reproduced on a modern base map at 1:50,000. In both cases the original OS Third Edition sheet margins and numbers are retained. The 1:250,000 sheets are defined using lines of latitude and longitude, each extending 1° north-south and 2° east-west.

Singapore

The first geological map of Singapore was produced in 1974, produced by the then Public Work Department. The publication includes a locality map, 8 map sheets detailing the topography and geological units, and a sheet containing cross sections of the island.

Since 1974, for 30 years, there were many findings reported in various technical conferences on new found geology islandwide, but no new publication was produced. In 2006, Defence Science & Technology Agency, with their developments in underground space promptly started a re-publication of the Geology of Singapore, second edition. The new edition that was published in 2009, contains a 1:75,000 geology map of the island, 6 maps (1:25,000) containing topography, street directory and geology, a sheet of cross section and a locality map.

The difference found between the 1976 Geology of Singapore report include numerous formations found in literature between 1976 and 2009. These include the Fort Canning Boulder Beds and stretches of limestone.

Web Mapping

Web map app in smart phone

Web mapping is the process of using maps delivered by geographic information systems (GIS). A web map on the World Wide Web is both served and consumed, thus web mapping is more than just web cartography, it is a service by which consumers may choose what the map will show. Web GIS emphasizes geodata processing aspects more involved with design aspects such as data acquisition and server software architecture such as data storage and algorithms, than it does the end-user reports themselves.

The terms *web GIS* and *web mapping* remain somewhat synonymous. Web GIS uses web maps, and end users who are *web mapping* are gaining analytical capabilities. The term *location-based services* refers to *web mapping* consumer goods and services. Web mapping usually involves a web browser or other user agent capable of client-server interactions. Questions of quality, usability, social benefits, and legal constraints are driving its evolution.

The advent of web mapping can be regarded as a major new trend in cartography. Until recently cartography was restricted to a few companies, institutes and mapping agencies, requiring relatively expensive and complex hardware and software as well as skilled cartographers and geomatics engineers.

Web mapping has brought many geographical datasets, including free ones generated by OpenStreetMap and proprietary datasets owned by Navteq, Google, Waze, and others. A range of free software to generate maps has also been conceived and implemented alongside proprietary tools like ArcGIS. As a result, the barrier to entry for serving maps on the web has been lowered.

Types of Web Maps

A first classification of web maps has been made by Kraak in 2001. He distinguished *static* and *dynamic* web maps and further distinguished *interactive* and *view only* web maps. Today there an increased number of dynamic web maps types, and static web map sources.

Analytical Web Maps

Analytical web maps offer GIS analysis. The geodata can be a static provision, or needs updates. The borderline between analytical web maps and web GIS is fuzzy. Parts of the analysis can be carried out by the GIS geodata server. As web clients gain capabilities processing is distributed.

Animated and Realtime

Realtime maps show the situation of a phenomenon in close to realtime (only a few seconds or minutes delay). They are usually animated. Data is collected by sensors and the maps are generated or updated at regular intervals or on demand.

Animated maps show changes in the map over time by animating one of the graphical or temporal variables. Technologies enabling client-side display of animated web maps include scalable vector graphics (SVG), Adobe Flash, Java, QuickTime, and others. Web maps with real-time animation include weather maps, traffic congestion maps and vehicle monitoring systems.

CartoDB launched an open source library, Torque, which enables the creation of dynamic animated maps with millions of records. Twitter uses this technology to create maps to reflect how users reacted to news and events worldwide.

Collaborative Web Maps

Collaborative maps are a developing potential. In proprietary or open source collaborative software, users collaborate to create and improve the web mapping experience. Some collaborative web mapping projects are:

- Google Map Maker
- Here Map Creator
- OpenStreetMap
- WikiMapia
- meta:Maps - a survey of Wikimedia web mapping proposals
- uebermaps - collaborative web mapping platform

Online Atlases

The traditional atlas goes through a remarkably large transition when hosted on the web. Atlases can cease their printed editions or offer printing on demand. Some atlases also offer raw data downloads of the underlying geospatial data sources.

Static Web Maps

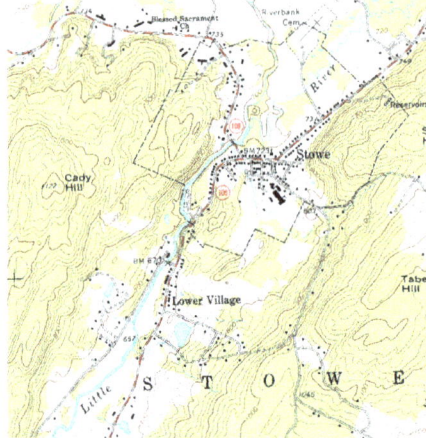

A USGS DRG - a static map

Static web pages are *view only* without animation or interactivity. These files are created once, often manually, and infrequently updated. Typical graphics formats for static web maps are PNG, JPEG, GIF, or TIFF (e.g., drg) for raster files, SVG, PDF or SWF for vector files. These include scanned paper maps not designed as screen maps. Paper maps have a much higher resolution and information density than typical computer displays of the same physical size, and might be unreadable when displayed on screens at the wrong resolution.

Evolving Paper Cartography

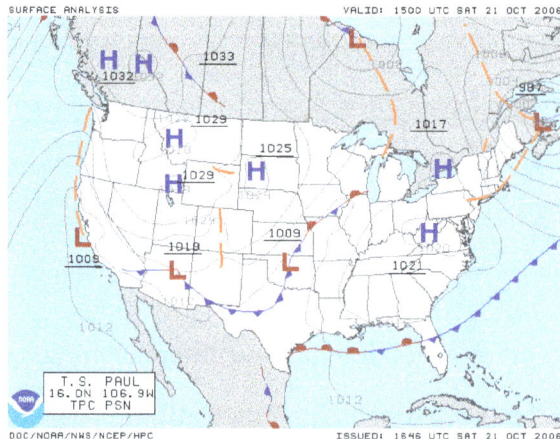

A surface weather analysis for the United States on October 21, 2006.

Compared to traditional techniques, mapping software has many advantages. The disadvantages are also stated.

- Web maps can easily *deliver up to date information*. If maps are generated automatically from databases, they can display information in almost realtime. They don't need to be printed, mastered and distributed. Examples:

    o   A map displaying election results, as soon as the election results become available.

    o   A traffic congestion map using traffic data collected by sensor networks.

    o   A map showing the current locations of mass transit vehicles such as buses or trains, allowing patrons to minimize their waiting time at stops or stations, or be aware of delays in service.

    o   Weather maps, such as NEXRAD.

- *Software and hardware infrastructure for web maps is cheap*. Web server hardware is cheaply available and many open source tools exist for producing web maps. Geodata, on the other hand, is not; satellites and fleets of automobiles use expensive equipment to collect the information on an ongoing basis. Perhaps owing to this, many people are still reluctant to publish geodata, especially in places where geodata are expensive. They fear copyright infringements by other people using their data without proper requests for permission.

- *Product updates can easily be distributed*. Because web maps distribute both logic and data with each request or loading, product updates can happen every time the web user reloads the application. In traditional cartography, when dealing with printed maps or interactive maps distributed on offline media (CD, DVD, etc.), a map update takes serious efforts, triggering a reprint or remastering as well as a redistribution of the media. With web maps, data and product updates are easier, cheaper, and faster, and occur more often. Perhaps owing to this, many web maps are of poor quality, both in symbolization, content and data accuracy.

- *Web maps can combine distributed data sources*. Using open standards and documented APIs one can integrate (*mash up*) different data sources, if the projection system, map scale and data quality match. The use of centralized data sources removes the burden for individual organizations to maintain copies of the same data sets. The downside is that one has to rely on and trust the external data sources. In addition, with detailed information available and the combination of distributed data sources, it is possible to find out and combine a lot of private and personal information of individual persons. Properties and estates of individuals are now accessible through high resolution aerial and satellite images throughout the world to anyone.

- *Web maps allow for personalization*. By using user profiles, personal filters and personal styling and symbolization, users can configure and design their own maps, if the web mapping systems supports personalization. Accessibility issues can be treated in the same way. If users can store their favourite colors and patterns they can avoid color combinations they can't easily distinguish (e.g. due to color blindness). Despite this, as with paper, web maps have the problem of limited screen space, but more so. This is in particular a problem for mobile web maps; the equipment carried usually has a very small screen, making it less likely that there is room for personalisation.

- *Web maps enable collaborative mapping* similar to web mapping technologies such as DHTML/Ajax, SVG, Java, Adobe Flash, etc. enable distributed data acquisition and collaborative efforts. Examples for such projects are the OpenStreetMap project or the Google Earth community. As with other open projects, quality assurance is very important, however, and the reliability of the internet and web server infrastructure is not yet good enough. Especially if a web map relies on external, distributed data sources, the original author often cannot guarantee the availability of the information.

- *Web maps support hyperlinking to other information on the web.* Just like any other web page or a wiki, web maps can act like an index to other information on the web. Any sensitive area in a map, a label text, etc. can provide hyperlinks to additional information. As an example a map showing public transport options can directly link to the corresponding section in the online train time table. However, development of web maps is complicated enough as it is: Despite the increasing availability of free and commercial tools to create web mapping and web GIS applications, it is still a more complex task to create interactive web maps than to typeset and print images. Many technologies, modules, services and data sources have to be mastered and integrated The development and debugging environments of a conglomerate of different web technologies is still awkward and uncomfortable.

History of Web Mapping

This section contains some of the milestones of web mapping, online mapping services and atlases.

- 1989: *Birth of the WWW*, WWW invented at CERN for the exchange of research documents.

- 1993: *Xerox PARC Map Viewer*, The first mapserver based on CGI/Perl, allowed reprojection styling and definition of map extent.

- 1994: *The World Wide Earthquake Locator*, the first interactive web mapping mashup was released, based on the Xerox PARC map view.

- 1994: *The National Atlas of Canada,* The first version of the National Atlas of Canada was released. Can be regarded as the first online atlas.

- 1995: *The Gazetteer for Scotland,* The prototype version of the Gazetteer for Scotland was released. The first geographical database with interactive mapping.

- 1995: *MapGuide,* First introduced as Argus MapGuide.

- 1996: Center for Advanced Spatial Technologies Interactive Mapper, Based on CGI/C shell/GRASS would allow the user to select a geographic extent, a raster base layer, and number of vector layers to create personalized map.

- 1996: *Mapquest,* The first popular online Address Matching and Routing Service with mapping output.

- 1996: *MultiMap,* The UK-based MultiMap website launched offering online mapping, routing and location based services. Grew into one of the most popular UK web sites.

- 1996: Geomedia WebMap 1.0, First version of Geomedia WebMap, already supports vector graphics through the use of ActiveCGM.

- 1996: *MapGuide,* Autodesk acquired Argus Technologies.and introduced Autodesk MapGuide 2.0.

National Atlas of the United States logo

- 1997: *US Online National Atlas Initiative,* The USGS received the mandate to coordinate and create the online National Atlas of the United States of America .

- 1997: UMN MapServer 1.0, Developed at the University of Minnesota (UMN) as Part of the NASA ForNet Project. Grew out of the need to deliver remote sensing data across the web for foresters.

- 1997: GeoInfoMapper - GeoInfo Solutions developed the first Java GIS Applet called 'JavaMap'. The application supported the export and conversion of MapInfo data for display in the thematic mapping tool for the web. GeoinfoMapper was demonstrated at the Victoria Computer Show in 1997 and referenced in the Universal Locator project at UC Berkeley School of Information.

- 1998: *Terraserver USA*, A Web Map Service serving aerial images (mainly b+w) and USGS DRGs was released. One of the first popular WMS. This service is a joint effort of USGS, Microsoft and HP.

- 1998: UMN MapServer 2.0, Added reprojection support (PROJ.4).

- 1998: MapObjects Internet Map Server, ESRI's entry into the web mapping business.

- 1999: *National Atlas of Canada, 6th edition*, This new version was launched at the ICA 1999 conference in Ottawa. Introduced many new features and topics. Is being improved gradually, since then, and kept up-to-date with technical advancements.

- 2000: ArcIMS 3.0, The first public release of ESRI's ArcIMS.

- 2000: ESRI Geography Network, ESRI founded Geography Network to distribute data and web map services.

- 2000: UMN MapServer 3.0, Developed as part of the NASA TerraSIP Project. This is also the first public, open source release of UMN Mapserver. Added raster support and support for TrueType fonts (FreeType).

- 2001: GeoServer, starts of the GeoServer project (Geoserver History)

- 2001: MapScript 1.0 for UMN MapServer, Adds a lot of flexibility to UMN MapServer solutions.

- 2001: *Tirolatlas*, A highly interactive online atlas, the first to be based on the SVG standard.

- 2002: UMN MapServer 3.5, Added support for PostGIS and ArcSDE. Version 3.6 adds initial OGC WMS support.

- 2002: ArcIMS 4.0, Version 4 of the ArcIMS web map server.

- 2003: *NASA World Wind*, NASA World Wind Released. An open virtual globe that loads data from distributed resources across the internet. Terrain and buildings can be viewed 3 dimensionally. The (XML based) markup language allows users to integrate their own personal content. This virtual globe needs special software and doesn't run in a web browser.

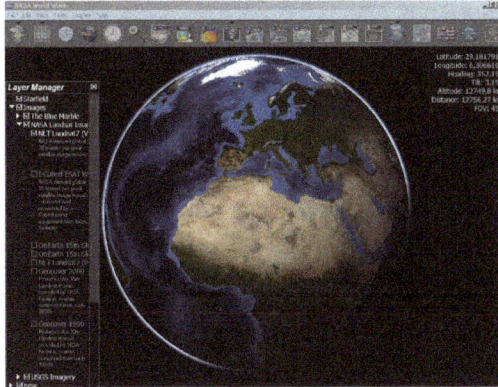

Screenshot from NASA World Wind

- 2003: UMN MapServer 4.0, Adds 24bit raster output support and support for PDF and SWF.

- 2004: OpenStreetMap, an open source, open content world map founded by Steve Coast.

- 2005: *Google Maps*, The first version of Google Maps. Based on raster tiles organized in a quad tree scheme, data loading done with XMLHttpRequests. This mapping application became highly popular on the web, also because it allowed other people to integrate google map services into their own website.

- 2005: *UMN MapServer* introduced as open source by the Open Source Geospatial Foundation (OSGeo). UMN MapServer 4.6, Adds support for SVG.

- 2005: *MapGuide Open Source* introduced as open source by Autodesk

- 2005: *Google Earth,* The first version of Google Earth was released building on the virtual globe metaphor. Terrain and buildings can be viewed 3 dimensionally. The KML (XML based) markup language allows users to integrate their own personal content. This virtual globe needs special software and doesn't run in a web browser.

- 2005: *OpenLayers*, the first version of the open source Javascript library Open-Layers.

- 2006: *WikiMapia* Launched

- 2009: Nokia makes *Ovi Maps* free on its smartphones.

- 2010: *MapBox* is founded

- 2012: Apple removes Google Maps as the default mapping app and replaces it with its own mapping app

- 2013: MapBox announces Vector Tiles for MapBox Streets

Web Mapping Technologies

Web mapping technologies require both server-side and client-side applications. The following is a list of technologies utilized in web mapping.

- Spatial databases are usually object relational databases enhanced with geographic data types, methods and properties. They are necessary whenever a web mapping application has to deal with dynamic data (that changes frequently) or with huge amount of geographic data. Spatial databases allow spatial queries, sub selects, reprojections, and geometry manipulations and offer various import and export formats. PostGIS is a prominent example; it is open source. MySQL also implements some spatial features. Oracle Spatial, Microsoft SQL Server (with the spatial extensions), and IBM DB2 are the commercial alternatives. The Open Geospacial Consortium's (OGC) specification "Simple Features" is a standard geometry data model and operator set for spatial databases. Part 2 of the specification defines an implementation using SQL.

- Tiled web maps display rendered maps made up of raster image "tiles".

- Vector tiles are also becoming more popular-- Google and Apple have both transitioned to vector tiles. Mapbox.com also offers vector tiles. This new style of web mapping is resolution independent, and also has the advantage of dynamically showing and hiding features depending on the interaction.

- WMS servers generate maps using parameters for user options such as the order of the layers, the styling and symbolization, the extent of the data, the data format, the projection, etc. The OGC standardized these options. Another WMS server standard is the Tile Map Service. Standard image formats include PNG, JPEG, GIF and SVG. Open source WMS Servers include UMN Mapserver, GeoServer and Mapnik. Commercial alternatives exist from most commercial GIS vendors, such as ESRI ArcIMS and CadCorp.

Scale (Map)

The scale of a map is the ratio of a distance on the map to the corresponding distance on the ground. This simple concept is complicated by the curvature of the Earth's surface, which forces scale to vary across a map. Because of this variation, the concept of scale becomes meaningful in two distinct ways. The first way is the ratio of the size of the generating globe to the size of the Earth. The generating globe is a conceptual model to which the Earth is shrunk and from which the map is projected.

The ratio of the Earth's size to the generating globe's size is called the nominal scale (= principal scale = representative fraction). Many maps state the nominal scale and may even display a bar scale (sometimes merely called a 'scale') to represent it. The second distinct concept of scale applies to the variation in scale across a map. It is the ratio of the mapped point's scale to the nominal scale. In this case 'scale' means the scale factor (= point scale = particular scale).

If the region of the map is small enough to ignore Earth's curvature—a town plan, for example—then a single value can be used as the scale without causing measurement errors. In maps covering larger areas, or the whole Earth, the map's scale may be less useful or even useless in measuring distances. The map projection becomes critical in understanding how scale varies throughout the map. When scale varies noticeably, it can be accounted for as the scale factor. Tissot's indicatrix is often used to illustrate the variation of point scale across a map.

The Terminology of Scales

Representation of Scale

Map scales may be expressed in words (a lexical scale), as a ratio, or as a fraction. Examples are:

'one centimetre to one hundred metres'   or   1:10,000   or   1/10,000

'one inch to one mile'   or   1:63,360   or   1/63,360

'one centimetre to one thousand kilometres'   or   1:100,000,000   or 1/100,000,000. (The ratio would usually be abbreviated to 1:100M)

Bar Scale vs. Lexical Scale

In addition to the above many maps carry one or more *(graphical)* bar scales. For example, some modern British maps have three bar scales, one each for kilometres, miles and nautical miles.

A lexical scale in a language known to the user may be easier to visualise than a ratio: if the scale is an inch to two miles and the map user can see two villages that are about two inches apart on the map, then it is easy to work out that the villages are about four miles apart on the ground.

A lexical scale may cause problems if it expressed in a language that the user does not understand or in obsolete or ill-defined units. For example, a scale of one inch to a furlong (1:7920) will be understood by many older people in countries where Imperial units used to be taught in schools. But a scale of one pouce to one league may be about 1:144,000, depending on the cartographer's choice of the many possible definitions for a league, and only a minority of modern users will be familiar with the units used.

Large Scale, Medium Scale, Small Scale

Contrast to spatial scale.

A map is classified as small scale or large scale or sometimes medium scale. Small scale refers to world maps or maps of large regions such as continents or large nations. In other words, they show large areas of land on a small space. They are called small scale because the representative fraction is relatively small.

Large scale maps show smaller areas in more detail, such as county maps or town plans might. Such maps are called large scale because the representative fraction is relatively large. For instance a town plan, which is a large scale map, might be on a scale of 1:10,000, whereas the world map, which is a small scale map, might be on a scale of 1:100,000,000.

The following table describes typical ranges for these scales but should not be considered authoritative because there is no standard:

Classification	Range	Examples
large scale	1:0 – 1:600,000	1:0.00001 for map of virus; 1:5,000 for walking map of town
medium scale	1:600,000 – 1:2,000,000	Map of a country
small scale	1:2,000,000 – 1:∞	1:50,000,000 for world map; $1:10^{21}$ for map of galaxy

The terms are sometimes used in the absolute sense of the table, but other times in a relative sense. For example, a map reader whose work refers solely to large-scale maps (as tabulated above) might refer to a map at 1:500,000 as small-scale.

In the English language, the word large-scale is often used to mean "extensive". However, as explained above, cartographers use the term "large scale" to refer to *less* extensive maps – those that show a smaller area. Maps that show an extensive area are "small scale" maps. This can be a cause for confusion.

Scale Variation

Mapping large areas causes noticeable distortions due to flattening the significantly curved surface of the earth. How distortion gets distributed depends on the map projection. Scale varies across the map, and the stated map scale will only be an approximation. This is discussed in detail below.

Large-scale Maps with Curvature Neglected

The region over which the earth can be regarded as flat depends on the accuracy of the survey measurements. If measured only to the nearest metre, then curvature of the earth is undetectable over a meridian distance of about 100 kilometres (62 mi) and over an east-west line of about 80 km (at a latitude of 45 degrees). If surveyed to the nearest 1 millimetre (0.039 in), then curvature is undetectable over a meridian distance of about 10 km and over an east-west line of about 8 km. Thus a plan of New York City accurate to one metre or a building site plan accurate to one millimetre would both satisfy the above conditions for the neglect of curvature. They can be treated by plane surveying and mapped by scale drawings in which any two points at the same distance on the drawing are at the same distance on the ground. True ground distances are calculated by measuring the distance on the map and then multiplying by the inverse of the scale fraction or, equivalently, simply using dividers to transfer the separation between the points on the map to a bar scale on the map.

Altitude Reduction

The variation in altitude, from the ground level down to the sphere's or ellipsoid's surface, also changes the scale of distance measurements.

Point Scale (or Particular Scale)

As proved by Gauss's *Theorema Egregium*, a sphere (or ellipsoid) cannot be projected onto a plane without distortion. This is commonly illustrated by the impossibility of smoothing an orange peel onto a flat surface without tearing and deforming it. The only true representation of a sphere at constant scale is another sphere such as a globe.

Given the limited practical size of globes, we must use maps for detailed mapping. Maps require projections. A projection implies distortion: A constant separation on the map does not correspond to a constant separation on the ground. While a map may display a graphical bar scale, the scale must be used with the understanding that it will be accurate on only some lines of the map.

Let P be a point at latitude ϕ and longitude λ on the sphere (or ellipsoid). Let Q be a neighbouring point and let α be the angle between the element PQ and the meridian at P: this angle is the azimuth angle of the element PQ. Let P' and Q' be corresponding points on the projection. The angle between the direction P'Q' and the projection of the meridian is the bearing β. In general $\alpha \neq \beta$. Comment: this precise distinction between azimuth (on the Earth's surface) and bearing (on the map) is not universally observed, many writers using the terms almost interchangeably.

Definition: the point scale at P is the ratio of the two distances P'Q' and PQ in the limit that Q approaches P. We write this as

$$\mu(\lambda,\phi,\alpha) = \lim_{Q \to P} \frac{P'Q'}{PQ},$$

where the notation indicates that the point scale is a function of the position of P and also the direction of the element PQ.

Definition: if P and Q lie on the same meridian $(\alpha = 0)$, the meridian scale is denoted by $h(\lambda,\phi)$.

Definition: if P and Q lie on the same parallel $(\alpha = \pi/2)$, the parallel scale is denoted by $k(\lambda,\phi)$.

Definition: if the point scale depends only on position, not on direction, we say that it is isotropic and conventionally denote its value in any direction by the parallel scale factor $k(\lambda,\phi)$.

Definition: A map projection is said to be conformal if the angle between a pair of lines intersecting at a point P is the same as the angle between the projected lines at the projected point P', for all pairs of lines intersecting at point P. A conformal map has an isotropic scale factor. Conversely isotropic scale factors across the map imply a conformal projection.

Isotropy of scale implies that *small* elements are stretched equally in all directions, that is the shape of a small element is preserved. This is the property of orthomorphism (from Greek 'right shape'). The qualification 'small' means that at some given accuracy of measurement no change can be detected in the scale factor over the element. Since conformal projections have an isotropic scale factor they have also been called orthomorphic projections. For example, the Mercator projection is conformal since it is constructed to preserve angles and its scale factor is isotopic, a function of latitude only: Mercator *does* preserve shape in small regions.

Definition: on a conformal projection with an isotropic scale, points which have the same scale value may be joined to form the isoscale lines. These are not plotted on maps for end users but they feature in many of the standard texts.

The Representative Fraction (RF) or Principal Scale

There are two conventions used in setting down the equations of any given projection. For example, the equirectangular cylindrical projection may be written as:

$$\text{cartographers:} \quad x = a\lambda \quad y = a\phi$$

$$\text{mathematicians:} \quad x = \lambda \quad y = \phi$$

Here we shall adopt the first of these conventions (following the usage in the surveys by Snyder). Clearly the above projection equations define positions on a huge cylinder wrapped around the Earth and then unrolled. We say that these coordinates define the projection map which must be distinguished logically from the actual printed (or viewed) maps. If the definition of point scale is in terms of the projection map then we can expect the scale factors to be close to unity. For normal tangent cylindrical projections the scale along the equator is k=1 and in general the scale changes as we move off the equator. Analysis of scale on the projection map is an investigation of the change of k away from its true value of unity.

Actual printed maps are produced from the projection map by a *constant* scaling denoted by a ratio such as 1:100M (for whole world maps) or 1:10000 (for such as town plans). To avoid confusion in the use of the word 'scale' this constant scale fraction is called the representative fraction (RF) of the printed map and it is to be identified with the ratio printed on the map. The actual printed map coordinates for the equirectangular cylindrical projection are

$$\text{printed map:} \qquad x = (RF)a\lambda \qquad y = (RF)a\phi$$

This convention allows a clear distinction of the intrinsic projection scaling and the reduction scaling.

From this point we ignore the RF and work with the projection map.

Visualisation of Point Scale: The Tissot Indicatrix

The Winkel tripel projection with Tissot's indicatrix of deformation

Consider a small circle on the surface of the Earth centred at a point P at latitude ϕ and longitude λ. Since the point scale varies with position and direction the projection of the circle on the projection will be distorted. Tissot proved that, as long as the distortion is not too great, the circle will become an ellipse on the projection. In general the dimension, shape and orientation of the ellipse will change over the projection. Superimposing these distortion ellipses on the map projection conveys the way in which the point scale is changing over the map. The distortion ellipse is known as Tissot's indicatrix. The example shown here is the Winkel tripel projection, the standard projection for world maps made by the National Geographic Society. The minimum distortion is on the central meridian at latitudes of 30 degrees (North and South).

Point Scale for Normal Cylindrical Projections of the Sphere

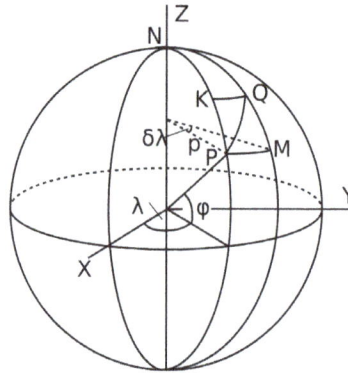

The key to a *quantitative* understanding of scale is to consider an infinitesimal element on the sphere. The figure shows a point P at latitude ϕ and longitude λ on the sphere. The point Q is at latitude $\phi + \delta\phi$ and longitude $\lambda + \delta\lambda$. The lines PK and MQ are arcs of meridians of length $a\delta\phi$ where a is the radius of the sphere and ϕ is in radian measure. The lines PM and KQ are arcs of parallel circles of length $(a\cos\phi)\delta\lambda$ with λ in radian measure. In deriving a *point* property of the projection *at* P it suffices to take an infinitesimal element PMQK of the surface: in the limit of Q approaching P such an element tends to an infinitesimally small planar rectangle.

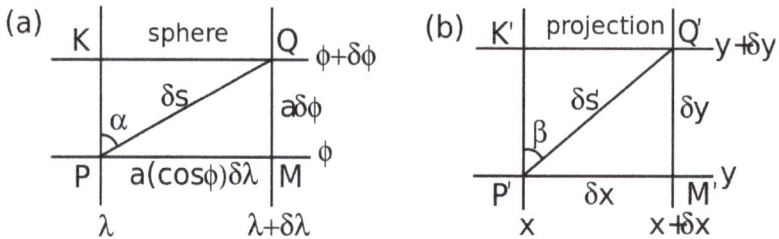

Infinitesimal elements on the sphere and a normal cylindrical projection

Normal cylindrical projections of the sphere have $x = a\lambda$ and y equal to a function of latitude only. Therefore, the infinitesimal element PMQK on the sphere projects to an infinitesimal element P'M'Q'K' which is an *exact* rectangle with a base $\delta x = a\delta\lambda$ and height δy. By comparing the elements on sphere and projection we can immediately deduce expressions for the scale factors on parallels and meridians. (The treatment of scale in a general direction may be found below.)

$$\text{parallel scale factor} \quad k = \frac{\delta x}{a\cos\phi\,\delta\lambda} = \sec\phi$$

$$\text{meridian scale factor} \quad h = \frac{\delta y}{a\delta\phi} = \frac{y'(\phi)}{a}$$

Note that the parallel scale factor $k = \sec\phi$ is independent of the definition of $y(\phi)$ so it is the same for all normal cylindrical projections. It is useful to note that at latitude 30 degrees the parallel scale is $k = \sec 30^\circ = 2/\sqrt{3} = 1.15$

at latitude 45 degrees the parallel scale is $k = \sec 45^\circ = \sqrt{2} = 1.414$

at latitude 60 degrees the parallel scale is $k = \sec 60^\circ = 2$

at latitude 80 degrees the parallel scale is $k = \sec 80^\circ = 5.76$

at latitude 85 degrees the parallel scale is $k = \sec 85^\circ = 11.5$

The following examples illustrate three normal cylindrical projections and in each case the variation of scale with position and direction is illustrated by the use of Tissot's indicatrix.

Three Examples of Normal Cylindrical Projection

The Equirectangular Projection

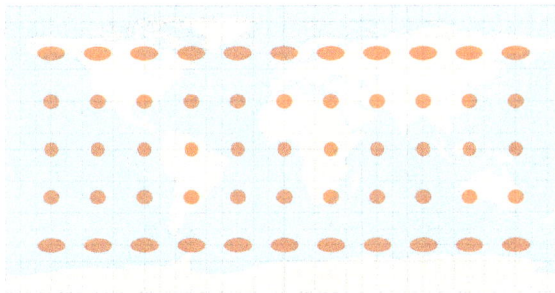

The equidistant projection with Tissot's indicatrix of deformation

The equirectangular projection, also known as the Plate Carrée (French for "flat square") or (somewhat misleadingly) the equidistant projection, is defined by

$$x = a\lambda, \quad y = a\phi,$$

where a is the radius of the sphere, λ is the longitude from the central meridian of the projection (here taken as the Greenwich meridian at $\lambda = 0$) and ϕ is the latitude. Note that λ and ϕ are in radians (obtained by multiplying the degree measure by a factor of $\pi/180$). The longitude λ is in the range $[-\pi, \pi]$ and the latitude ϕ is in the range $[-\pi/2, \pi/2]$.

Since $y'(\phi) = 1$

parallel scale, $k = \dfrac{\delta x}{a\cos\phi\,\delta\lambda} = \sec\phi$

meridian scale $h = \dfrac{\delta y}{a\delta\phi} = 1$

The figure illustrates the Tissot indicatrix for this projection. On the equator h=k=1 and the circular elements are undistorted on projection. At higher latitudes the circles are distorted into an ellipse given by stretching in the parallel direction only: there is no distortion in the meridian direction. The ratio of the major axis to the minor axis is $\sec\phi$. Clearly the area of the ellipse increases by the same factor.

It is instructive to consider the use of bar scales that might appear on a printed version of this projection. The scale is true (k=1) on the equator so that multiplying its length on a printed map by the inverse of the RF (or principal scale) gives the actual circumference of the Earth. The bar scale on the map is also drawn at the true scale so that transferring a separation between two points on the equator to the bar scale will give the correct distance between those points. The same is true on the meridians. On a parallel other than the equator the scale is $\sec\phi$ so when we transfer a separation from a parallel to the bar scale we must divide the bar scale distance by this factor to obtain the distance between the points when measured along the parallel (which is not the true distance along a great circle). On a line at a bearing of say 45 degrees ($\beta = 45°$) the scale is continuously varying with latitude and transferring a separation along the line to the bar scale does not give a distance related to the true distance in any simple way. Even if we could work out a distance along this line of constant bearing its relevance is questionable since such a line on the projection corresponds to a complicated curve on the sphere. For these reasons bar scales on small-scale maps must be used with extreme caution.

Mercator Projection

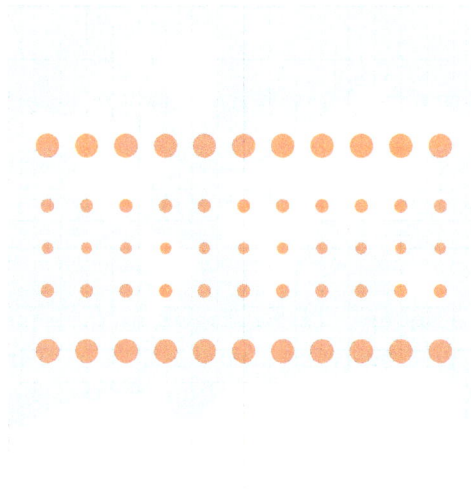

The Mercator projection with Tissot's indicatrix of deformation. (The distortion increases without limit at higher latitudes)

The Mercator projection maps the sphere to a rectangle (of infinite extent in the y-direction) by the equations

$$x = a\lambda$$

$$y = a\ln\left[\tan\left(\frac{\pi}{4}+\frac{\phi}{2}\right)\right]$$

where a, λ and ϕ are as in the previous example. Since $y'(\phi) = a\sec\phi$ the scale factors are:

$$\text{parallel scale} \quad k = \frac{\delta x}{a\cos\phi\,\delta\lambda} = \sec\phi.$$

$$\text{meridian scale} \quad h = \frac{\delta y}{a\delta\phi} = \sec\phi.$$

In the mathematical addendum it is shown that the point scale in an arbitrary direction is also equal to $\sec\phi$ so the scale is isotropic (same in all directions), its magnitude increasing with latitude as $\sec\phi$. In the Tissot diagram each infinitesimal circular element preserves its shape but is enlarged more and more as the latitude increases.

Lambert's Equal Area Projection

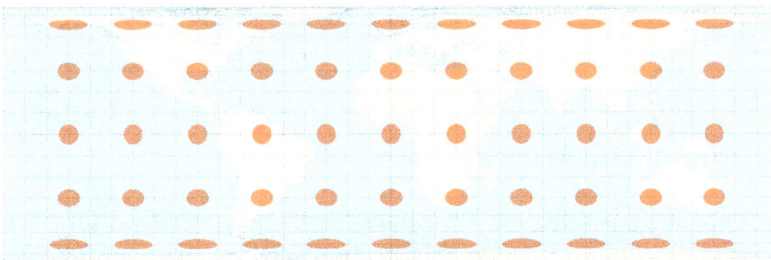

Lambert's normal cylindrical equal-area projection with Tissot's indicatrix of deformation

Lambert's equal area projection maps the sphere to a finite rectangle by the equations

$$x = a\lambda \qquad\qquad y = a\sin\phi$$

where a, λ and ϕ are as in the previous example. Since $y'(\phi) = \cos\phi$ the scale factors are

$$\text{parallel scale} \quad k = \frac{\delta x}{a\cos\phi\,\delta\lambda} = \sec\phi$$

$$\text{meridian scale} \quad h = \frac{\delta y}{a\delta\phi} = \cos\phi$$

The calculation of the point scale in an arbitrary direction is given below.

The vertical and horizontal scales now compensate each other (hk=1) and in the Tissot diagram each infinitesimal circular element is distorted into an ellipse of the *same* area as the undistorted circles on the equator.

Graphs of Scale Factors

The graph shows the variation of the scale factors for the above three examples. The top plot shows the isotropic Mercator scale function: the scale on the parallel is the same as the scale on the meridian. The other plots show the meridian scale factor for the Equirectangular projection (h=1) and for the Lambert equal area projection. These last two projections have a parallel scale identical to that of the Mercator plot. For the Lambert note that the parallel scale (as Mercator A) increases with latitude and the meridian scale (C) decreases with latitude in such a way that hk=1, guaranteeing area conservation.

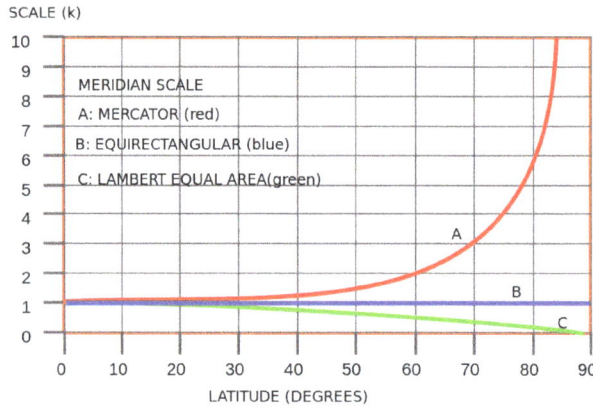

Scale Variation on the Mercator Projection

The Mercator point scale is unity on the equator because it is such that the auxiliary cylinder used in its construction is tangential to the Earth at the equator. For this reason the usual projection should be called a tangent projection. The scale varies with latitude as $k = \sec \phi$. Since $\sec \phi$ tends to infinity as we approach the poles the Mercator map is grossly distorted at high latitudes and for this reason the projection is totally inappropriate for world maps (unless we are discussing navigation and rhumb lines). However, at a latitude of about 25 degrees the value of $\sec \phi$ is about 1.1 so Mercator *is* accurate to within 10% in a strip of width 50 degrees centred on the equator. Narrower strips are better: a strip of width 16 degrees (centred on the equator) is accurate to within 1% or 1 part in 100.

A standard criterion for good large-scale maps is that the accuracy should be within 4 parts in 10,000, or 0.04%, corresponding to $k = 1.0004$. Since $\sec \phi$ attains this value at $\phi = 1.62$ degrees. Therefore, the tangent Mercator projection is highly accurate within a strip of width 3.24 degrees centred on the equator. This corresponds to north-south distance of about 360 km (220 mi). Within this strip Mercator is *very* good, highly accurate and shape preserving because it is conformal (angle preserving). These observations prompted the development of the transverse Mercator projections in which a meridian is treated 'like an equator' of the projection so that we obtain an accurate map within a narrow distance of that meridian. Such maps are good for countries aligned nearly north-south (like Great Britain) and a set of 60 such maps is used for the Universal Transverse Mercator (UTM). Note that in both these projections (which are based on various ellipsoids) the transformation equations for x and y and the expression for the scale factor are complicated functions of both latitude and longitude.

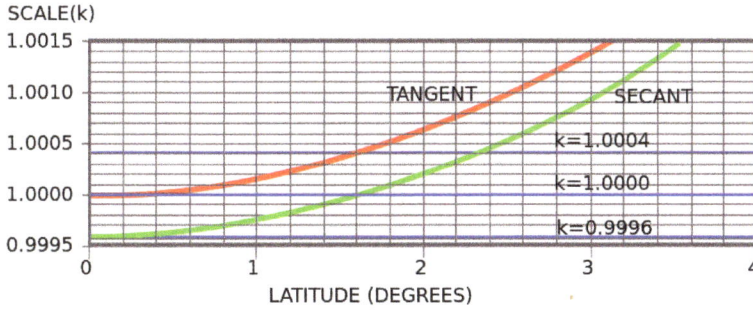

Scale variation near the equator for the tangent (red) and secant (green) Mercator projections.

Secant, or Modified, Projections

The basic idea of a secant projection is that the sphere is projected to a cylinder which intersects the sphere at two parallels, say ϕ_1 north and south. Clearly the scale is now true at these latitudes whereas parallels beneath these latitudes are contracted by the projection and their (parallel) scale factor must be less than one. The result is that deviation of the scale from unity is reduced over a wider range of latitudes.

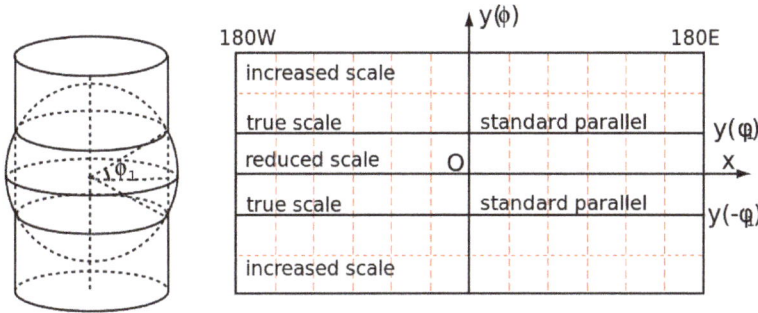

As an example, one possible secant Mercator projection is defined by

$$x = 0.9996a\lambda \qquad y = 0.9996a \ln\left(\tan\left(\frac{\pi}{4} + \frac{\phi}{2} \right) \right).$$

The numeric multipliers do not alter the shape of the projection but it does mean that the scale factors are modified:

secant Mercator scale, $\quad k = 0.9996\sec\phi.$

Thus

- the scale on the equator is 0.9996,

- the scale is k=1 at a latitude given by ϕ_1 where $\sec\phi_1 = 1/0.9996 = 1.00004$ so that $\phi = 1.62$ degrees,

- k=1.0004 at a latitude ϕ_2 given by $\sec\phi_2 = 1.0004/0.9996 = 1.0008$ for which $\phi_2 = 2.29$ degrees. Therefore, the projection has $1 < k < 1.0004$, that is an accuracy of 0.04%, over a wider strip of 4.58 degrees (compared with 3.24 degrees for the tangent form).

This is illustrated by the lower (green) curve in the figure.

Such narrow zones of high accuracy are used in the UTM and the British OSGB projection, both of which are secant, transverse Mercator on the ellipsoid with the scale on the central meridian constant at $k_0 = 0.9996$. The isoscale lines with $k = 1$ are slightly curved lines approximately 180 km east and west of the central meridian. The maximum value of the scale factor is 1.001 for UTM and 1.0007 for OSGB.

The lines of unit scale at latitude ϕ_1 (north and south), where the cylindrical projection surface intersects the sphere, are the standard parallels of the secant projection.

Whilst a narrow band with $|k-1| < 0.0004$ is important for high accuracy mapping at a large scale, for world maps much wider spaced standard parallels are used to control the scale variation. Examples are

- Behrmann with standard parallels at 30N, 30S.

- Gall equal area with standard parallels at 45N, 45S.

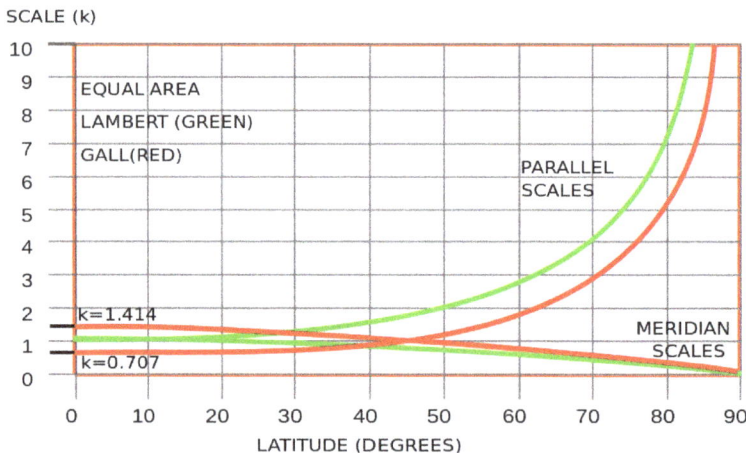

Scale variation for the Lambert (green) and Gall (red) equal area projections.

The scale plots for the latter are shown below compared with the Lambert equal area scale factors. In the latter the equator is a single standard parallel and the parallel scale increases from k=1 to compensate the decrease in the meridian scale. For the Gall the parallel scale is reduced at the equator (to k=0.707) whilst the meridian scale is increased (to k=1.414). This gives rise to the gross distortion of shape in the Gall-Peters projection. (On the globe Africa is about as long as it is broad). Note that the meridian and parallel scales are both unity on the standard parallels.

Mathematical Addendum

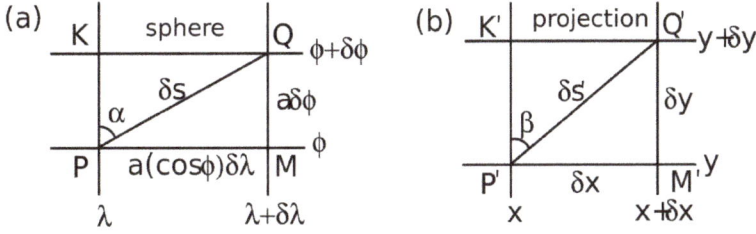

Infinitesimal elements on the sphere and a normal cylindrical projection

For normal cylindrical projections the geometry of the infinitesimal elements gives

$$\text{(a)} \quad \tan \alpha = \frac{a \cos \phi \, \delta \lambda}{a \delta \phi},$$

$$\text{(b)} \quad \tan \beta = \frac{\delta x}{\delta y} = \frac{a \delta \lambda}{\delta y}.$$

The relationship between the angles β and α is

$$\text{(c)} \quad \tan \beta = \frac{a \sec \phi}{y'(\phi)} \tan \alpha.$$

For the Mercator projection $y'(\phi) = a \sec \phi$ giving $\alpha = \beta$: angles are preserved. (Hardly surprising since this is the relation used to derive Mercator). For the equidistant and Lambert projections we have $y'(\phi) = a$ and $y'(\phi) = a \cos \phi$ respectively so the relationship between α and β depends upon the latitude ϕ. Denote the point scale at P when the infinitesimal element PQ makes an angle α with the meridian by μ_α. It is given by the ratio of distances:

$$\mu_\alpha = \lim_{Q \to P} \frac{P'Q'}{PQ} = \lim_{Q \to P} \frac{\sqrt{\delta x^2 + \delta y^2}}{\sqrt{a^2 \, \delta \phi^2 + a^2 \cos^2 \phi \delta \lambda^2}}.$$

Setting $\delta x = a \delta \lambda$ and substituting δx and $\delta \lambda$ from equations (a) and (b) respectively gives

$$\mu_\alpha(\phi) = \sec \phi \left[\frac{\sin \alpha}{\sin \beta} \right].$$

For the projections other than Mercator we must first calculate β from α and ϕ using equation (c), before we can find μ_α. For example, the equirectangular projection has $y' = a$ so that

$$\tan \beta = \sec \phi \tan \alpha.$$

If we consider a line of constant slope β on the projection both the corresponding value of α and the scale factor along the line are complicated functions of ϕ. There is no simple way of transferring a general finite separation to a bar scale and obtaining meaningful results.

Map and Globe

The word globe comes from the Latin word globus which means round mass or sphere. A terrestrial globe is a three dimensional scaled model of Earth. The fact that earth resembles a sphere was established by Greek astronomy in the third century BC and the earliest globes came from that period. Unlike maps, the globe is a representation which is free from distortions (distortion in shape, area). The modern globes have longitudes and latitudes marked over it so that one is able to tell the approximate coordinates of a specific place. To make the illustration better, people have tried depicting variations of earth surface over the globe. The relief raised globes allow a user to visualize the mountain ranges and plains as the features are modeled using elevations and depressions. But the relief is not scaled rather exaggerated. The raised relief would be virtually invisible if a scale representation were attempted.

A Globe depicting relief

The map and the globe are similar in a manner that both of them represent earth (on particular scales) but there also exist a few differences between them, which are enumerated below:

Globe	Map
Three dimensional representation of earth in the form of a sphere	Two dimensional representation of earth in the form of a flat surface
Impossible to see all the countries of the world at a glance as only half of the globe can be seen at a time	All countries of the world can be seen on a world map at a glance

The shape and size of geographical features is correctly represented.	Due to projection there are distortions in shape and size of geographical features.
Accurate tracing of the maps is not possible due to the curvature of the globe	Maps can be accurately traced
A part of earth can't be separately represented on the globe	A part of earth can be separately represented on the map
Inconvenient to carry	Easy to carry

Map Projection

Map projection is a mathematical expression using which the three-dimensional surface of earth is represented in a two dimensional plane. The process of projection results in distortion of one or more map properties such as shape, size, area or direction.

A single projection system can never account for the correct representation of all map properties for all the regions of the world. Therefore, hundreds of projection systems have been defined for accurate representation of a particular map element for a particular region of the world.

Classification of Map Projections

Map projections are classified on the following criteria:

- Method of construction

- Development surface used

- Projection properties

- Position of light source

Method of Construction

The term map projection implies projecting the graticule of the earth onto a flat surface with the help of shadow cast. However, not all of the map projections are developed in this manner. Some projections are developed using mathematical calculations only. Given below are the projections that are based on the method of construction:

Perspective Projections: These projections are made with the help of shadow cast from an illuminated globe on to a developable surface

Non Perspective Projections: These projections do not use shadow cast from an illuminated globe on to a developable surface. A developable surface is only assumed to be covering the globe and the construction of projections is done using mathematical calculations.

Development Surface

Projection transforms the coordinates of earth on to a surface that can be flattened to a plane without distortion (shearing or stretching). Such a surface is called a developable surface. The three basic projections are based on the types of developable surface and are introduced below:

1. Cylindrical Projection

- It can be visualized as a cylinder wrapped around the globe.

- Once the graticule is projected onto the cylinder, the cylinder is opened to get a grid like pattern of latitudes and longitudes.

- The longitudes (meridians) and latitudes (parallels) appear as straight lines

- Length of equator on the cylinder is equal to the length of the equator therefore is suitable for showing equatorial regions.

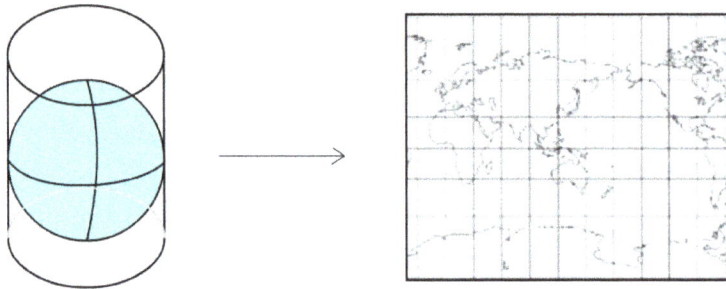

Aspects of cylindrical projection:

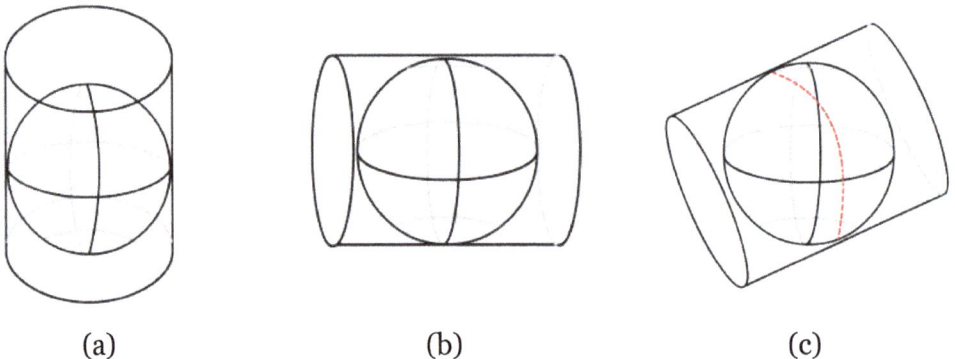

| (a)   (b)   (c) |

(a) Normal: when cylinder has line of tangency to the equator. It includes Equirectangular Projection, the Mercator projection, Lambert's Cylindrical Equal Area, Gall's Stereographic Cylindrical, and Miller cylindrical projection.

(b) Transverse: when cylinder has line of tangency to the meridian. It includes the Cassini Projection, Transverse Mercator, Transverse cylindrical Equal Area Projection, and Modified Transverse Mercator.

(c) Oblique: when cylinder has line of tangency to another point on the globe. It only consists of the Oblique Mercator projection.

2. Conic Projection

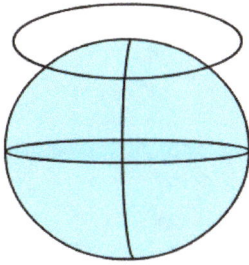

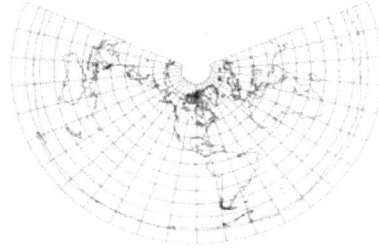

- It can be visualized as a cone placed on the globe, tangent to it at some parallel.

- After projecting the graticule on to the cone, the cone is cut along one of the meridian and unfolded. Parallels appear as arcs with a pole and meridians as straight lines that converge to the same point.

- It can represent only one hemisphere, at a time, northern or southern.

- Suitable for representing middle latitudes.

Aspects of conic projection:

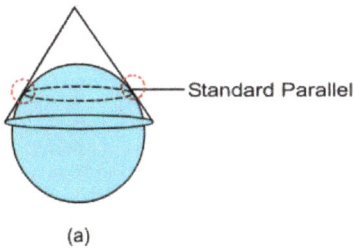

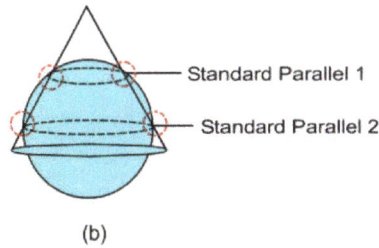

(a)   (b)

(a) Tangent: when the cone is tangent to only one of the parallel.

(b) Secant: when the cone is not big enough to cover the curvature of earth, it intersects the earth twice at two parallels.

3. Azimuthal/Zenithal Projection

- It can be visualized as a flat sheet of paper tangent to any point on the globe

- The sheet will have the tangent point as the centre of the circular map, where meridians passing through the centre are straight line and the parallels are seen as concentric circle.

- Suitable for showing polar areas

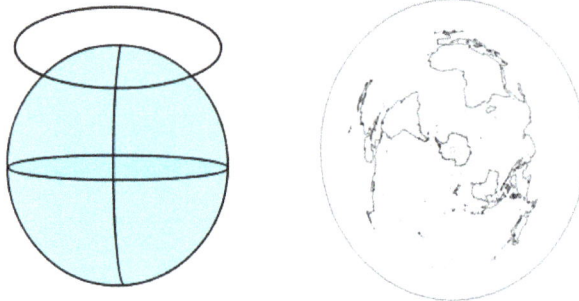

Aspects of zenithal projection:

(a) Equatorial zenithal: When the plane is tangent to a point on the equator.

(b) Oblique zenithal: when the plane is tangent to a point between a pole and the equator.

(c) Polar zenithal: when the plane is tangent to one of the poles.

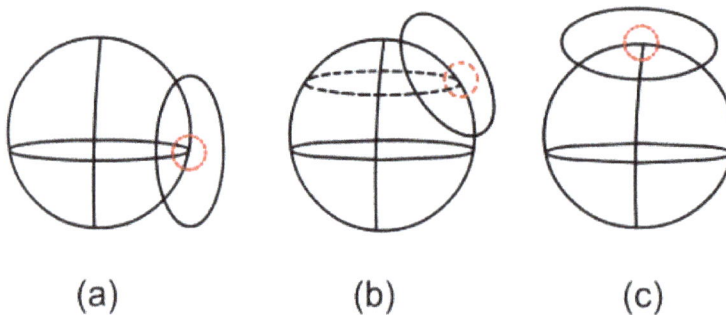

(a)   (b)   (c)

Projection Properties

According to properties map projections can be classified as:

Equal area projection: Also known as homolographic projections. The areas of different parts of earth are correctly represented by such projections.

True shape projection: Also known as orthomorphic projections. The shapes of different parts of earth are correctly represented on these projections.

True scale or equidistant projections: Projections that maintain correct scale are called true scale projections. However, no projection can maintain the correct scale throughout. Correct scale can only be maintained along some parallel or meridian.

Position of Light Source

Placing light source illuminating the globe at different positions results in the development of different projections. These projections are:

Gnomonic projection: when the source of light is placed at the centre of the globe

Stereographic Projection: when the source of light is placed at the periphery of the globe, diametrically opposite to the point at which developable surface touches the globe

Orthographic Projection: when the source of light is placed at infinity from the globe opposite to the point at which developable surface touches the globe

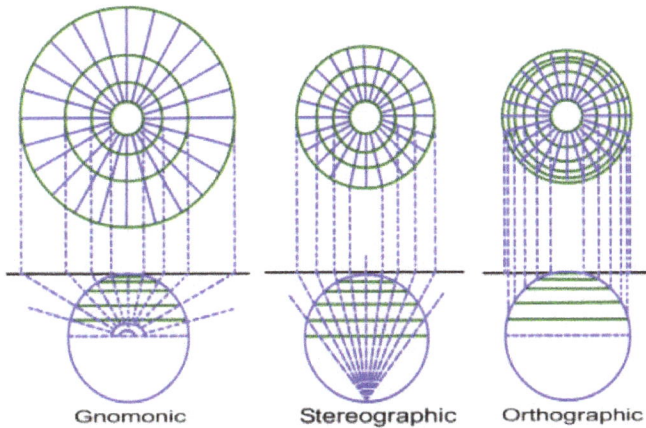

Gnomonic   Stereographic   Orthographic

Projections and position of light source

Constructing Map Projections

Cylindrical Projection

Let us draw a network of Simple cylindrical Projection for the whole globe on the scale of 1: 400,000,000 spacing meridians and parallels at 30° interval

Calculations:

Radius of the earth=635,000,000 cm

Radius of globe on 1:400,000,000=1/400000000×635000000

$$=1.587 \ cm$$

Length of the equator on the globe=$2\pi r$, where r is the radius of globe

$$= 2 \times 22 / 7 \times 1.587$$

$$= 9.975 \, cm$$

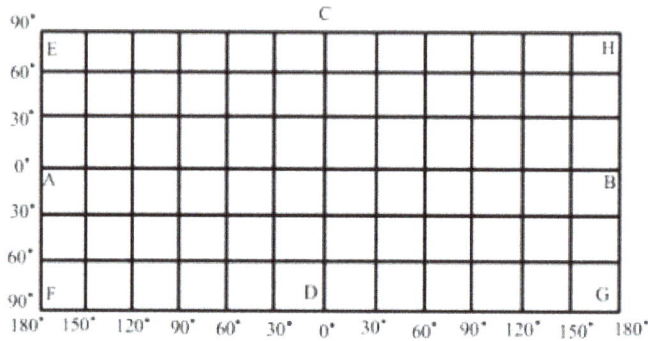

Simple cylindrical projection graticule

Steps of construction:

- Draw a line AB, 9.975 cm long to represent the equator. The equator is a circle on the globe and is subtended by 360°.

- Since the meridians are to be drawn at an interval of 30° divide AB into 360/30 or 12 equal parts.

- The length of a meridian is equal to half the length of the equator i.e. 9.975/2 or 4.987 cm.

- To draw meridians, erect perpendiculars on the points of divisions of AB. Take these perpendiculars equal to the length specified for a meridian and keep half of their length on either side of the equator.

- A meridian on a globe is subtended by 180°. Since the parallels are to be drawn at an interval of 30°, divide the central meridian into 180/30 i.e. 6 parts.

- Through these points of divisions draw lines parallel to the equator. These lines will be parallels of latitude. Mark the equator and the central meridian with 0° and the parallels and other meridians. EFGH is the required graticule.

Conical Projection

Let us draw a graticule on simple conical projection with one standard parallel on the scale of 1: 180,000,000 for the area extending from the equator to 90° N latitude and from 60° W longitude to 100° E longitude with parallels spaced at 15° interval, meridians at 20°, and standard parallel 45° N.

Calculations:

Radius of the earth=635,000,000 cm

Therefore, radius of the globe on the scale of 1:180,000,000

$$=1/180,000,000 \times 635,000,000$$

$$=3.527 \ cm$$

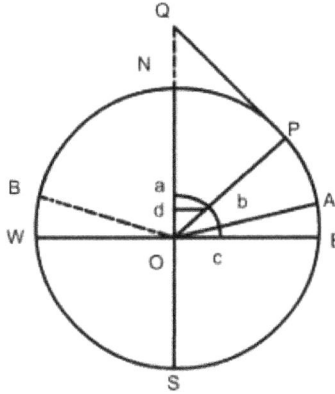

Steps of construction:

1. Draw a circle with a radius of 3.527 cm that represents the globe. Let NS be the polar diameter and WE be the equatorial diameter which intersect each other at right angles at O.

2. To draw the standard parallel 45° N, draw OP making an angle of 45° with OE.

3. Draw QP tangent to OP and extend ON to meet PQ at point Q.

4. Draw OA making an angle equal to the parallel interval i.e. 15° with OE.

Draw line LM, it represents the central meridian

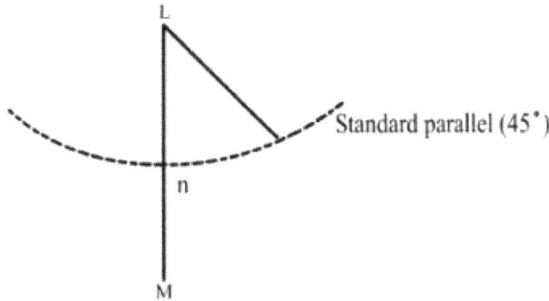

With L as the centre and QP as the radius, draw an arc intersecting LM at n. This arc describes the standard parallel 45° N.

The distance between the successive parallels is 15°. The length of the arc subtended by 15° is calculated as under:

$$2\pi r \times 1/360 \times 15$$
$$= 2 \times 22/7 \times 3.527 \times 15/360$$
$$= 0.923\,\text{cm}$$

From point n, mark off distances nr, rs, st, nu, uv and vM , each distance being equal to 0.923 cm. With L as centre, draw arcs passing through the points t, s, r, u, v and M. These arcs represent the parallels.

5. Draw OB making an angle of 20° with OW Length of the arc subtended by 20° is calculated as under:

$$2\pi r \times 1/360 \times 20$$
$$= 2 \times 22/7 \times 3.527 \times 20/360$$
$$= 1.231 cm$$

•   With O as centre and radius equal to the arc WB (1.231 cm) draw arc abc.

•   From point b, drop perpendicular bd on line ON. Now db is the distance between the meridians.

Keeping in view the number of meridians to be drawn, mark off distances along the standard parallel toward the east and west of the point n, each distance being equal to db.

Join point L with the points of divisions marked on the standard parallel and produce them to meet the equator.

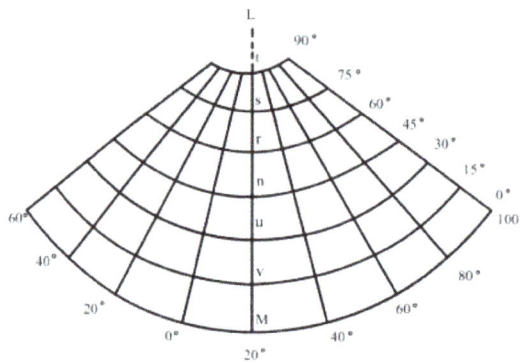

Simple conic projection

Azimuthal Projection

Let us draw Polar zenithal equal area projection for the northern hemisphere on the scale of 1: 200,000,000 spacing parallels at 15° interval and meridians at 30° interval.

Calculations:

Radius of the earth=635,000,000 cm

Therefore, radius of the globe on the scale of 1:200,000,000

$$=1/200,000,000 \times 635,000,000$$
$$=3.175 \ cm$$

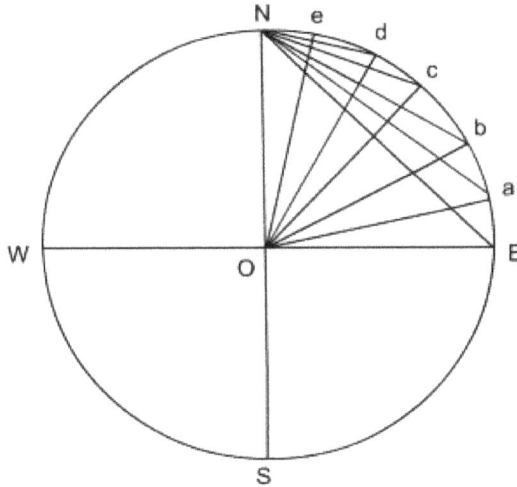

Steps of construction:

- Draw a circle with radius equal to 3.175 cm representing a globe. Let NS and WE be the polar and equatorial diameter respectively which intersect each other at right angles at O, the centre of the circle.

- Draw radii Oa, Ob, Oc, Od, and Oe making angles of 15°, 30°, 45°, 60° and 75° respectively with OE. Join Ne, Nd, Nc, Nb, Na and NE by straight lines.

- With radius equal to Ne, and N' as centre draw a circle. This circle represents 75° parallel. Similarly with centre N' and radii equal to Nd, Nc, Nb, Na and NE draw circles to represent the parallels of 60°, 45°, 30°, 15° and 0° respectively.

- Draw straight lines AB and CD intersecting each other at the centre i.e. point N.

- Radius N'B represents 0° meridian, N'A 180° meridian, N'D 90° E meridian and N'C 90° W meridian.

- Using protractor, draw other radii at 30° interval to represent other meridians

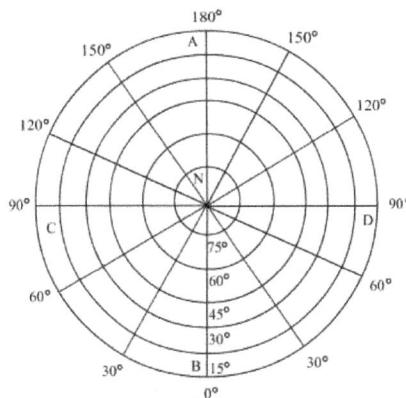

Polar zenithal equal area projection

Selection of Map Projection

Choosing a correct map projection for an area depends on the following:

Map Purpose

Considering the purpose of the map is important while choosing the map projection. If a map has a specific purpose, one may need to preserve a certain property such as shape, area or direction

On the basis of the property preserved, maps can be categorized as following

a) Maps that preserve shapes.

Used for showing local directions and representing the shapes of the features. Such maps include:

- Topographic and cadastral maps.
- Navigation charts (for plotting course bearings and wind direction).
- Civil engineering maps and military maps.
- Weather maps (for showing the local direction in which weather systems are moving).

b) Maps that preserve area

The size of any area on the map is in true proportion to its size on the earth. Such projections can be used to show

- Density of an attribute e.g. population density with dots
- Spatial extent of a categorical attribute e.g. land use maps
- Quantitative attributes by area e.g. Gross Domestic Product by country
- World political maps to correct popular misconceptions about the relative sizes of countries.

c) Maps that preserve scale

Preserves true scale from a single point to all other points on the map. The maps that use this property include:

- Maps of airline distances from a single city to several other cities
- Seismic maps showing distances from the epicenter of an earthquake
- Maps used to calculate ranges; for example, the cruising ranges of airplanes or the habitats of animal species

d) Maps that preserve direction

On any Azimuthal projection, all azimuths, or directions, are true from a single spec-ified point to all other points on the map. On a conformal projection, directions are locally true, but are distorted with distance.

General Purpose Maps

There are many projections which show the world with a balanced distortion of shape and area. Few of these are Winkel Tripel, Robinson and Miller Cylindrical.For larg-er-scale maps, from continents to large countries, equidistant projections are good at balancing shape and area distortion. Depending on the area of interest, one might use Azimuthal Equidistant,Equidistant Conic and Plate Carrée.

Study Area

Geographical Location

The line of zero distortion for a cylindrical projection is equator. For conical projections it is parallels and for Azimuthal it is one of the poles. If the study area is in tropics use cylindrical projection, for middle latitudes use conical and for Polar Regions use Azi-muthal projections.

Shape of the Area

Young in 1920 described a way of selecting the map projection which is known as Young's rule. According to this rule, if the ratio of maximum extent (z) (measured from the centre of the country to its most distant boundary) and the width (δ) of the country comes out to be less than 1.41, Azimuthal projection is preferable. If the ratio is greater than 1.41 a conical or cylindrical projection should be used.

Z/δ < 1.41 Azimuthal Projection

Z/δ >1.41 Conical or Cylindrical projections

Projection Systems

Given below is the description of the projection systems that are mostly used:

Cylindrical Projection

I.   Equirectangular projection

This is a Projection on to a cylinder which is tangent to the equator. It is believed to be invented by Marinus of Tyre, about A.D. 100.

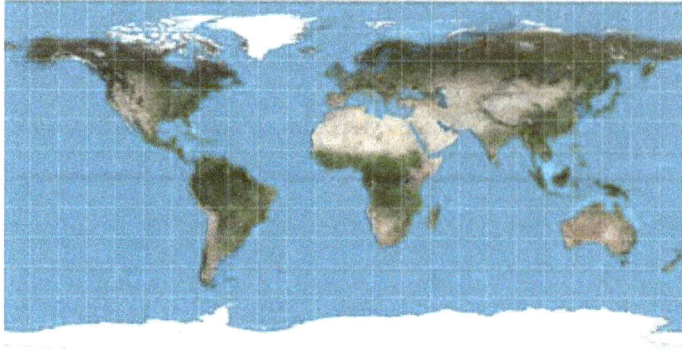

1. Poles are straight lines equal in length to the equator

2. Meridians are straight parallel lines, equally spaced and are half as long as the equator. All meridians are of same length therefore scale is true along all meridians.

3. Parallels are straight, equally spaced lines which are perpendicular to the meridians and are equal to the length of the equator.

4. Length of the equator on the map is the same as that on the globe but the length of other parallels on map is more than the length of corresponding parallels on the globe. So the scale is true only along the Equator and not along other parallels.

5. Distance between the parallels and meridians remain same throughout the map.

6. Since the projection is neither equal area nor orthomorphic, maps on this projection are used for general purposes only.

II. Lambert's cylindrical equal-area projection

It is devised by JH Lambert in 1772. It is a normal perspective projection onto a cylinder tangent at the equator

1. Parallels and meridians are straight lines

2. The meridians intersect parallels at right angles

3. The distance between parallels decrease toward the poles but meridians are equally spaced

4. The length of the equator on this projection is same as that on globe but other parallels are longer than corresponding parallels on globe. So, the scale is true along the equator but is exaggerated along other parallels

5. Shape and scale distortions increase near points 90 degrees from the central line resulting in vertical exaggeration of Equatorial regions with compression of regions in middle latitudes

6. Despite the shape distortion in some portions of a world map, this projection is well suited for equal-area mapping of regions which are predominantly north-south in extent, which have an oblique central line, or which lie near the Equator.

III. Gall's stereographic cylindrical projection

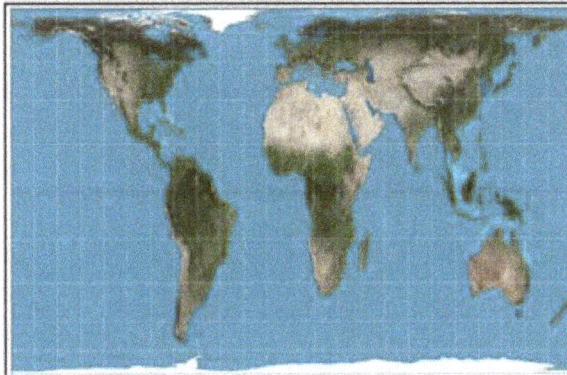

Invented by James Gall in 1855, this projection is a cylindrical projection with two standard parallels at 45°N and 45°S.

- Poles are straight lines.

- Meridians are straight lines and are equally spaced.

- Parallels are straight lines but the distance between them increases away from the equator.

- Shapes are true at the standard parallels. Distortion increases on moving away from these latitudes and is highest at the poles.

- Scale is true in all directions along 45°N and 45°S.

- Used for world maps in British atlases.

IV. Mercator projection

Gerardus Mercator in 1569 invented this projection.

- Parallels and meridians are straight lines

- Meridians intersect parallels at right angle

- Distance between the meridians remains the same but distance between the parallels increases towards the pole

- The length of equator on the projection is equal to the length of the equator on the globe whereas other parallels are drawn longer than what they are on the globe, therefore the scale along the equator is correct but is incorrect for other parallels

- As scale varies from parallel to parallel and is exaggerated towards the pole, the shapes of large sized countries are distorted more towards pole and less towards equator. However, shapes of small countries are preserved

- The image of the poles are at infinity

- Commonly used for navigational purposes, ocean currents and wind direction are shown on this projection

V. Transverse Mercator

This projection results from projecting the sphere onto a cylinder tangent to a central meridian.

- Only centre meridian and equator are projected as straight lines. The other parallels and meridians are projected as curves.

- The meridians and the parallels intersect at right angles

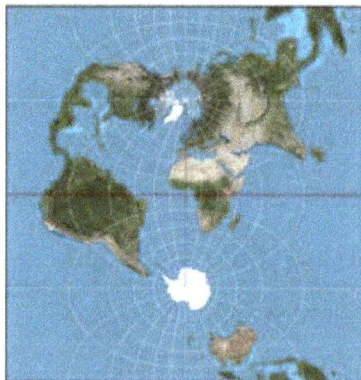

- Small shapes are maintained but larger shapes distort away from the central meridian.

- The area distortion increases with distance from the central meridian

- Used to portray areas with larger north-south extent. British National Grid is based on this projection only.

Pseudo-cylindrical Projections

A pseudo cylindrical projection is that projection in which latitudes are parallel straight lines but meridians are curved.

I.   Mollweide Projection

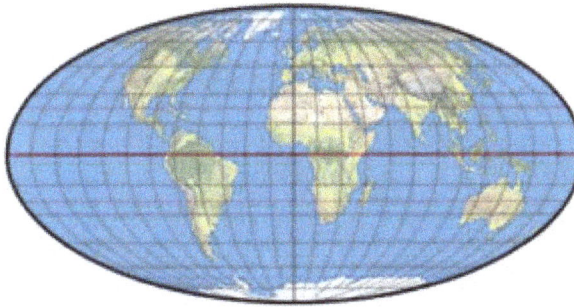

- The poles are points and the central meridian is a straight line

- The meridians 90° away from central meridians are circular arcs and all other meridians are elliptical arcs.

- The parallels are straight but unequally spaced.

- Scale is true along 40° 44' North and 40° 44' South.

- Equal –area projection

- Used for preparing world maps

II. Sinusoidal Projection

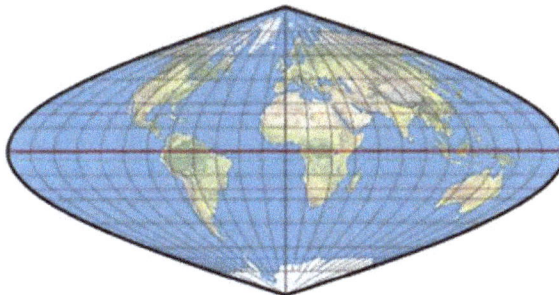

- The central meridian is a straight line and all other meridians are equally spaced sinusoidal curves.

- The parallels are straight lines that intersect centre meridian at right angles.

- Shape and angles are correct along the central meridian and equator

- The distortion of shape and angles increases away from the central meridian and is high near the edges

- Equal area projection

- Used for world maps illustrating area characteristics. Used for continental maps of South America, Africa, and occasionally other land masses, where each has its own central meridian.

III. Eckert VI

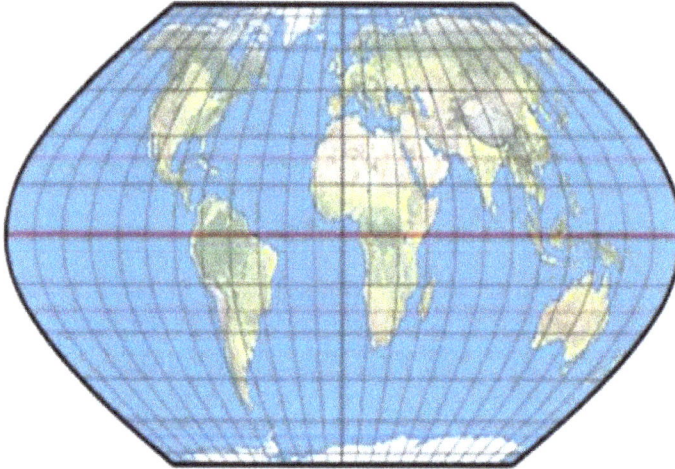

- Parallels are unequally spaced straight lines.

- Meridians are equally spaced sinusoidal curves.

- The poles and the central meridians are straight lines and half as long as equator.

- It stretches shapes and scale by 29% in the north-south direction, along the equator. This stretching reduces to zero at 49° 16' N and 49° 16' S.

- The areas near the poles are compressed in north-south direction.

- Suitable for thematic mapping of the world.

Conical Projection

I.   Bonne's Projection

- Pole is represented as a point and parallels as concentric arcs of circles

- Scale along all the parallels is correct

- Central meridian is a straight line along which the scale is correct.

- Other meridians are curved and longer than corresponding meridians on the globe. Scale along meridians increases away from the central meridian

- Central meridian intersects all parallels at right angle. Other meridians intersect standard parallel at right angle but other parallels obliquely. Shape is only preserved along central meridian and standard parallel

- The distance and scale between two parallels are correct. Area between projected parallels is equal to the area between the same parallels on the globe. Therefore, is an equal area projection

- Maps of European countries are shown in this projection. It is also used for preparing topographical sheets of small countries of middle latitudes.

II. Polyonic Projection

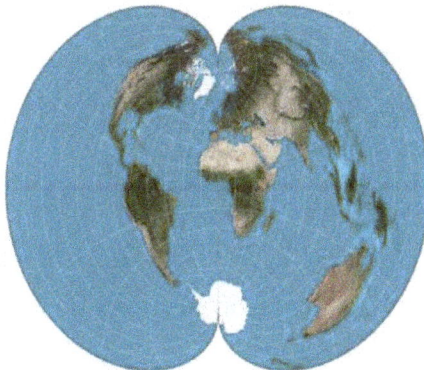

- The parallels are arcs of circles with different centers

- Each parallel is a standard parallel i.e. each parallel is developed from a different cone

- Equator is represented as a straight line and the pole as a point

- Parallels are equally spaced along central meridian but the distance between them increases away from the central meridian.

- Scale is correct along every parallel.

- Central meridian intersects all parallels at right angle so the scale along it, is correct. Other meridians are curved and longer than corresponding meridians on the globe and so scale along meridians increases away from the central meridian.

- It is used for preparing topographical sheets of small areas.

III. Gnomonic Projection

It is also known as great-circle sailing chart.

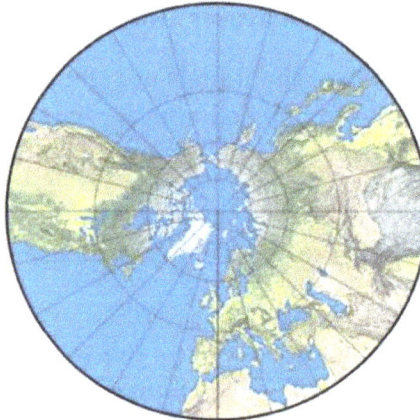

- The pole is a point forming the centre of the projection and the parallels are concentric circles.

- The meridians are straight lines radiating from pole having correct angular distance between them.

- The meridians intersect the parallels at right angles.

- The parallels are unequally spaced. The distances between the parallels increase rapidly toward the margin of the projection. This causes exaggeration of the scale along the meridians.

- The scale along the parallels increases away from the centre of the projection.

- The exaggeration and distortion of shapes increases away from the centre of the projection. The exaggeration in the meridian scale is greater than that in any other zenithal projection.

- It is neither equal area nor orthomorphic.

- An arc on the globe which is a part of a great circle is represented as a straight line on this projection. This is because the radii from the centre of the globe are produced to meet the plane placed tangentially at the pole.

- It is used to show great-circle paths as straight lines and thus to assist navigators and aviators in determining appropriate courses.

IV. Sereographic Projection

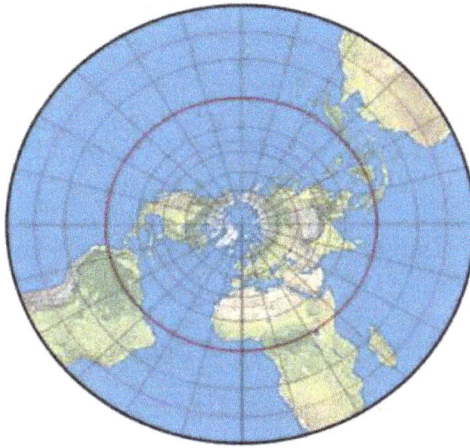

- The pole is a point forming the centre of the projection and the parallels are concentric circles.

- The meridians are straight lines radiating from pole having correct angular distance between them.

- The meridians intersect the parallels at right angles.

- The parallels are unequally spaced. The distances between the parallels increase toward the margin of the projection. The exaggeration in the meridian scale is less than that in the case of Gnomonic projection.

- The scale along the parallels also increases away from the meridian and in the same proportion in which it increases along the meridians. At any point scale along the parallel is equal to the scale along the meridian.

- The areas are exaggerated on this projection and the exaggeration increases away from the centre of the projection.

- A circle drawn on the globe is represented by a circle on this projection.

- It is used to show world in hemispheres. Also used for preparing aeronautical charts and daily weather maps of the polar areas.

V. Orthographic Projection

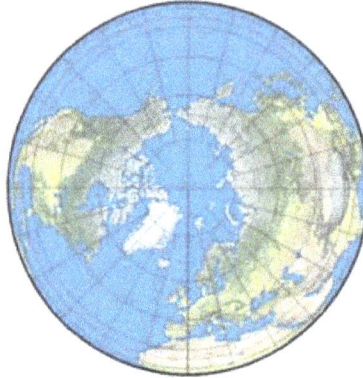

- The pole is a point forming the centre of the projection and the parallels are concentric circles.

- The meridians are straight lines radiating from pole having correct angular distance between them.

- The meridians intersect the parallels at right angles.

- The parallels are not equally spaced. The distances between them decrease rapidly towards the margin of the projection. So, the scale along the meridians decreases away from the centre of the projection.

- The scale along the parallel is correct.

- The distortion of the shapes increases away from the centre of the projection.

- It is neither equal area nor orthomorphic.

- The projection is used to prepare charts for showing the celestial bodies such as moon and other planets.

Universal Transverse Mercator (UTM)

UTM projection divides the surface of the Earth into a number of zones, each zone having a 6 degree longitudinal extent, Transverse Mercator projection with a central meridian in the center of the zone. UTM zones extend from 80 degrees South latitude to 84 degrees North latitude. The zones are numbered from west to east. The first zone begins at the International Date Line (180°).

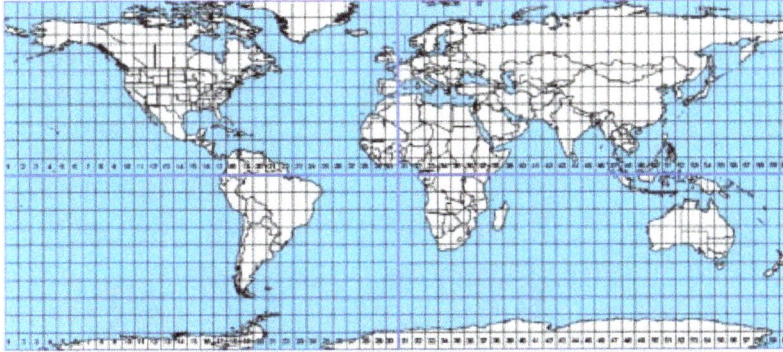

The particular transverse Mercator map that is used to represent each zone has its central meridian running north-to-south down the center of the zone. This means that no portion of any particular zone is very far from the central meridian of the transverse Mercator map that is used to depict the zone. Since a Universal Transverse Mercator zone is 6° of longitude wide, no portion of a UTM zone is more than 3° of longitude from the zone's central meridian. Since the distortion in a transverse Mercator map is relatively low near the map's central meridian, the result of this close proximity to the map's central meridian is that the transverse Mercator map used to depict each zone within the coordinate system contains relatively little distortion.

Adapted from the report Map Projections of Europe (2001), the table gives an account of the commonly used projection systems.

Property	Developable surface	Aspect	Projections	Extent of use
Conformal (True shape)	Cylinder	Normal	Mercator	Equatorial regions (east-west extent)
		Transverse	UTM (Universal Transverse Mercator)	Whole world except polar areas
		Oblique	Rosenmund Oblique Mercator	Small regions, oblique & east - west extent
	Cone	Normal	Lambert Conformal Conic	Small regions, oblique & east - west extent (1 or 2 standard parallels)
	Plane	Any	Stereographic	Small regions upto the hemisphere
		Polar	UPS (Universal Polar Stereographic)	Polar regions
Homolographic (Equal area)	Cylinder	Normal	Lambert Equal Area	Equatorial areas (east-west extent)
	Cone	Normal	Albers Equal Area	Smaller regions & continents with east-west extent
	Plane	Any	Lambert Azimuthal Equal Area	Smaller regions about same north-south , east-west extent
		Equatorial	Hammer-Aitoff	World

Equidistant	Cylinder	Normal	Plate Caree	World
		Transverse	Cassini Soldner	Locally used for large scale mapping
	Cone	Normal	Equidistant Conic	Smaller regions & continents with (1 or 2 standard parallels) east-west extent
	Plane	Any	Azimuthal Equidistant	Smaller regions about same north-south , east-west extent

A medieval depiction of the Ecumene (1482, Johannes Schnitzer, engraver), constructed after the coordinates in Ptolemy's *Geography* and using his second map projection

A map projection is a systematic transformation of the latitudes and longitudes of locations on the surface of a sphere or an ellipsoid into locations on a plane. Maps cannot be created without map projections. All map projections necessarily distort the surface in some fashion. Depending on the purpose of the map, some distortions are acceptable and others are not; therefore, different map projections exist in order to preserve some properties of the sphere-like body at the expense of other properties. There is no limit to the number of possible map projections.

More generally, the surfaces of planetary bodies can be mapped even if they are too irregular to be modeled well with a sphere or ellipsoid. Even more generally, projections are a subject of several pure mathematical fields, including differential geometry, projective geometry, and manifolds. However, "map projection" refers specifically to a cartographic projection.

Background

Maps can be more useful than globes in many situations: they are more compact and easier to store; they readily accommodate an enormous range of scales; they are viewed easily on computer displays; they can facilitate measuring properties of the terrain being mapped; they can show larger portions of the Earth's surface at once; and they are cheaper to produce and transport. These useful traits of maps motivate the development of map projections.

However, Carl Friedrich Gauss's Theorema Egregium proved that a sphere's surface cannot be represented on a plane without distortion. The same applies to other reference surfaces used as models for the Earth. Since any map projection is a representation of one of those surfaces on a plane, all map projections distort. Every distinct map projection distorts in a distinct way. The study of map projections is the characterization of these distortions.

Projection is not limited to perspective projections, such as those resulting from casting a shadow on a screen, or the rectilinear image produced by a pinhole camera on a flat film plate. Rather, any mathematical function transforming coordinates from the curved surface to the plane is a projection. Few projections in actual use are perspective.

For simplicity, most of this article assumes that the surface to be mapped is that of a sphere. In reality, the Earth and other large celestial bodies are generally better modeled as oblate spheroids, whereas small objects such as asteroids often have irregular shapes. These other surfaces can be mapped as well. Therefore, more generally, a map projection is any method of "flattening" a continuous curved surface onto a plane.

Metric Properties of Maps

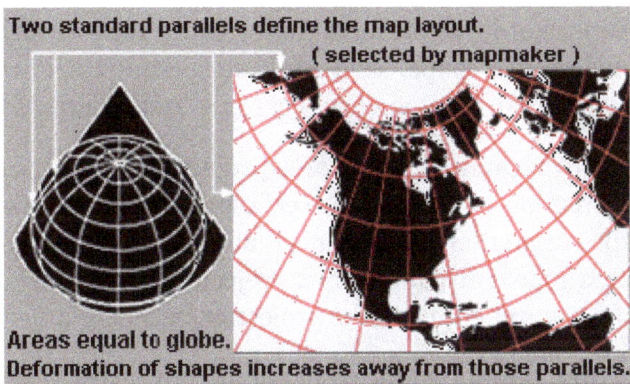

An Albers projection shows areas accurately, but distorts shapes.

Many properties can be measured on the Earth's surface independent of its geography. Some of these properties are:

- Area
- Shape
- Direction
- Bearing
- Distance
- Scale

Map projections can be constructed to preserve at least one of these properties, though only in a limited way for most. Each projection preserves or compromises or approximates basic metric properties in different ways. The purpose of the map determines which projection should form the base for the map. Because many purposes exist for maps, a diversity of projections have been created to suit those purposes.

Another consideration in the configuration of a projection is its compatibility with data sets to be used on the map. Data sets are geographic information; their collection depends on the chosen datum (model) of the Earth. Different datums assign slightly different coordinates to the same location, so in large scale maps, such as those from national mapping systems, it is important to match the datum to the projection. The slight differences in coordinate assignation between different datums is not a concern for world maps or other vast territories, where such differences get shrunk to imperceptibility.

Which Projection is Best?

The mathematics of projection do not permit any particular map projection to be "best" for everything. Something will always be distorted. Therefore, many projections exist to serve the many uses of maps and their vast range of scales.

Modern national mapping systems typically employ a transverse Mercator or close variant for large-scale maps in order to preserve conformality and low variation in scale over small areas. For smaller-scale maps, such as those spanning continents or the entire world, many projections are in common use according to their fitness for the purpose.

Thematic maps normally require an equal area projection so that phenomena per unit area are shown in correct proportion. However, representing area ratios correctly necessarily distorts shapes more than many maps that are not equal-area. Hence reference maps of the world often appear on compromise projections instead. Due to distortions inherent in any map of the world, the choice of projection becomes largely one of aesthetics.

The Mercator projection, developed for navigational purposes, has often been used in world maps where other projections would have been more appropriate. This problem has long been recognized even outside professional circles. For example, a 1943 *New York Times* editorial states:

The time has come to discard [the Mercator] for something that represents the continents and directions less deceptively... Although its usage... has diminished... it is still highly popular as a wall map apparently in part because, as a rectangular map, it fills a rectangular wall space with more map, and clearly because its familiarity breeds more popularity.

A controversy in the 1980s over the Peters map motivated the American Cartographic Association (now Cartography and Geographic Information Society) to produce a series of booklets (including *Which Map Is Best*) designed to educate the public about map projections and distortion in maps. In 1989 and 1990, after some internal debate, seven North American geographic organizations adopted a resolution recommending against using any rectangular projection (including Mercator and Gall–Peters) for reference maps of the world.

Distortion

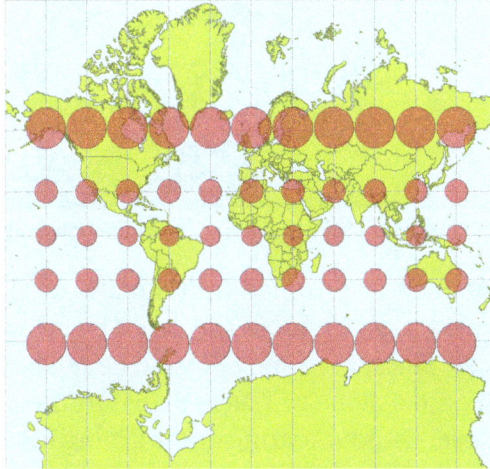

Tissot's Indicatrices on the Mercator projection

The classical way of showing the distortion inherent in a projection is to use Tissot's indicatrix. For a given point, using the scale factor h along the meridian, the scale factor k along the parallel, and the angle θ' between them, Nicolas Tissot described how to construct an ellipse that characterizes the amount and orientation of the components of distortion. By spacing the ellipses regularly along the meridians and parallels, the network of indicatrices shows how distortion varies across the map.

Construction of a Map Projection

The creation of a map projection involves two steps:

1. Selection of a model for the shape of the Earth or planetary body (usually choosing between a sphere or ellipsoid). Because the Earth's actual shape is irregular, information is lost in this step.

2. Transformation of geographic coordinates (longitude and latitude) to Cartesian (x,y) or polar plane coordinates. In large-scale maps, cartesian coordinates normally have a simple relation to eastings and northings defined as a grid superimposed on the projection. In small-scale maps, eastings and northings are not meaningful, and grids are not superimposed.

Some of the simplest map projections are literal projections, as obtained by placing a light source at some definite point relative to the globe and projecting its features onto a specified surface. This is not the case for most projections, which are defined only in terms of mathematical formulae that have no direct geometric interpretation.

Choosing a Projection Surface

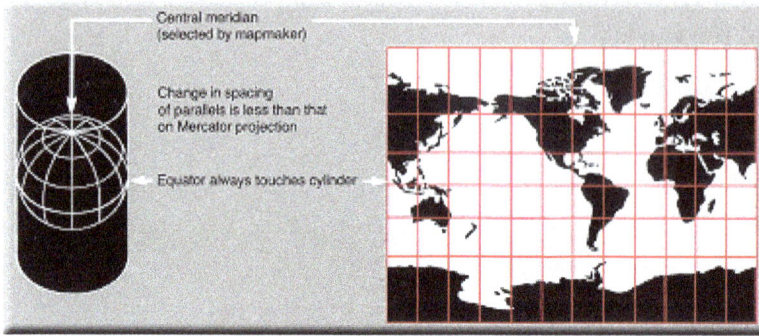

A Miller cylindrical projection maps the globe onto a cylinder.

A surface that can be unfolded or unrolled into a plane or sheet without stretching, tearing or shrinking is called a *developable surface.* The cylinder, cone and the plane are all developable surfaces. The sphere and ellipsoid do not have developable surfaces, so any projection of them onto a plane will have to distort the image. (To compare, one cannot flatten an orange peel without tearing and warping it.)

One way of describing a projection is first to project from the Earth's surface to a developable surface such as a cylinder or cone, and then to unroll the surface into a plane. While the first step inevitably distorts some properties of the globe, the developable surface can then be unfolded without further distortion.

Aspect of the Projection

Once a choice is made between projecting onto a cylinder, cone, or plane, the aspect of the shape must be specified. The aspect describes how the developable surface is placed relative to the globe: it may be *normal* (such that the surface's axis of symmetry coincides with the Earth's axis), *transverse* (at right angles to the Earth's axis) or *oblique* (any angle in between).

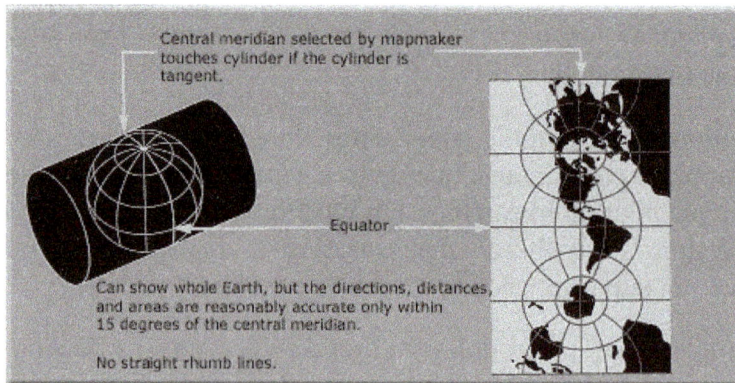

This transverse Mercator projection is mathematically the same as a standard
Mercator, but oriented around a different axis.

Notable Lines

The developable surface may also be either *tangent* or *secant* to the sphere or ellipsoid. Tangent means the surface touches but does not slice through the globe; secant means the surface does slice through the globe. Moving the developable surface away from contact with the globe never preserves or optimizes metric properties, so that possibility is not discussed further here.

Tangent and secant lines (*standard lines*) are represented undistorted. If these lines are a parallel of latitude, as in conical projections, it is called a *standard parallel*. The *central meridian* is the meridian to which the globe is rotated before projecting. The central meridian (usually written λ_o) and a parallel of origin (usually written φ_o) are often used to define the origin of the map projection.

Scale

A globe is the only way to represent the earth with constant scale throughout the entire map in all directions. A map cannot achieve that property for any area, no matter how small. It can, however, achieve constant scale along specific lines.

Some possible properties are:

- The scale depends on location, but not on direction. This is equivalent to preservation of angles, the defining characteristic of a conformal map.

- Scale is constant along any parallel in the direction of the parallel. This applies for any cylindrical or pseudocylindrical projection in normal aspect.

- Combination of the above: the scale depends on latitude only, not on longitude or direction. This applies for the Mercator projection in normal aspect.

- Scale is constant along all straight lines radiating from a particular geographic location. This is the defining characteristic of an equidistant projection such as the Azimuthal equidistant projection. There are also projections (Maurer, Close) where true distances from *two* points are preserved.

Choosing a Model for the Shape of the Body

Projection construction is also affected by how the shape of the Earth or planetary body is approximated. The earth is taken as a sphere in order to simplify the discussion. However, the Earth's actual shape is closer to an oblate ellipsoid. Whether spherical or ellipsoidal, the principles discussed hold without loss of generality.

Selecting a model for a shape of the Earth involves choosing between the advantages and disadvantages of a sphere versus an ellipsoid. Spherical models are useful for small-scale maps such as world atlases and globes, since the error at that scale is not usually noticeable or important enough to justify using the more complicated ellipsoid. The ellipsoidal model is commonly used to construct topographic maps and for other large- and medium-scale maps that need to accurately depict the land surface. Auxiliary latitudes are often employed in projecting the ellipsoid.

A third model is the geoid, a more complex and accurate representation of Earth's shape coincident with what mean sea level would be if there were no winds, tides, or land. Compared to the best fitting ellipsoid, a geoidal model would change the characterization of important properties such as distance, conformality and equivalence. Therefore, in geoidal projections that preserve such properties, the mapped graticule would deviate from a mapped ellipsoid's graticule. Normally the geoid is not used as an Earth model for projections, however, because Earth's shape is very regular, with the undulation of the geoid amounting to less than 100 m from the ellipsoidal model out of the 6.3 million m Earth radius. For irregular planetary bodies such as asteroids, however, sometimes models analogous to the geoid are used to project maps from.

Classification

A fundamental projection classification is based on the type of projection surface onto which the globe is conceptually projected. The projections are described in terms of placing a gigantic surface in contact with the earth, followed by an implied scaling operation. These surfaces are cylindrical (e.g. Mercator), conic (e.g. Albers), or azimuthal or plane (e.g. stereographic). Many mathematical projections, however, do not neatly fit into any of these three conceptual projection methods. Hence other peer categories have been described in the literature, such as pseudoconic, pseudocylindrical, pseudoazimuthal, retroazimuthal, and polyconic.

Another way to classify projections is according to properties of the model they preserve. Some of the more common categories are:

- Preserving direction (*azimuthal or zenithal*), a trait possible only from one or two points to every other point

- Preserving shape locally (*conformal* or *orthomorphic*)

- Preserving area (*equal-area* or *equiareal* or *equivalent* or *authalic*)

- Preserving distance (*equidistant*), a trait possible only between one or two points and every other point

- Preserving shortest route, a trait preserved only by the gnomonic projection

Because the sphere is not a developable surface, it is impossible to construct a map projection that is both equal-area and conformal.

Projections by Surface

The three developable surfaces (plane, cylinder, cone) provide useful models for understanding, describing, and developing map projections. However, these models are limited in two fundamental ways. For one thing, most world projections in actual use do not fall into any of those categories. For another thing, even most projections that do fall into those categories are not naturally attainable through physical projection. As L.P. Lee notes,

> No reference has been made in the above definitions to cylinders, cones or planes. The projections are termed cylindric or conic because they can be regarded as developed on a cylinder or a cone, as the case may be, but it is as well to dispense with picturing cylinders and cones, since they have given rise to much misunderstanding. Particularly is this so with regard to the conic projections with two standard parallels: they may be regarded as developed on cones, but they are cones which bear no simple relationship to the sphere. In reality, cylinders and cones provide us with convenient descriptive terms, but little else.

Lee's objection refers to the way the terms *cylindrical*, *conic*, and *planar* (azimuthal) have been abstracted in the field of map projections. If maps were projected as in light shining through a globe onto a developable surface, then the spacing of parallels would follow a very limited set of possibilities. Such a cylindrical projection (for example) is one which:

1. Is rectangular;

2. Has straight vertical meridians, spaced evenly;

3. Has straight parallels symmetrically placed about the equator;

4. Has parallels constrained to where they fall when light shines through the globe onto the cylinder, with the light source someplace along the line formed by the intersection of the prime meridian with the equator, and the center of the sphere.

(If you rotate the globe before projecting then the parallels and meridians will not necessarily still be straight lines. Rotations are normally ignored for the purpose of classification.)

Where the light source emanates along the line described in this last constraint is what yields the differences between the various "natural" cylindrical projections. But the term *cylindrical* as used in the field of map projections relaxes the last constraint entirely. Instead the parallels can be placed according to any algorithm the designer has decided suits the needs of the map. The famous Mercator projection is one in which the placement of parallels does not arise by "projection"; instead parallels are placed how they need to be in order to satisfy the property that a course of constant bearing is always plotted as a straight line.

Cylindrical

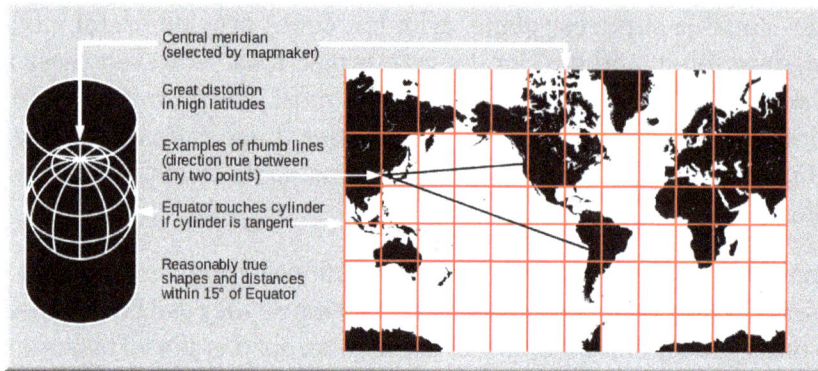

The Mercator projection shows rhumbs as straight lines. A rhumb is a course of constant bearing. Bearing is the compass direction of movement.

The term "normal cylindrical projection" is used to refer to any projection in which meridians are mapped to equally spaced vertical lines and circles of latitude (parallels) are mapped to horizontal lines.

The mapping of meridians to vertical lines can be visualized by imagining a cylinder whose axis coincides with the Earth's axis of rotation. This cylinder is wrapped around the Earth, projected onto, and then unrolled.

By the geometry of their construction, cylindrical projections stretch distances east-west. The amount of stretch is the same at any chosen latitude on all cylindrical projections, and is given by the secant of the latitude as a multiple of the equator's scale. The various cylindrical projections are distinguished from each other solely by their north-south stretching (where latitude is given by φ):

- North-south stretching equals east-west stretching (sec φ): The east-west scale matches the north-south scale: conformal cylindrical or Mercator; this distorts areas excessively in high latitudes.

- North-south stretching grows with latitude faster than east-west stretching (sec^2 φ): The cylindric perspective (or central cylindrical) projection; unsuitable because distortion is even worse than in the Mercator projection.

- North-south stretching grows with latitude, but less quickly than the east-west stretching: such as the Miller cylindrical projection $(sec4/5\varphi)$.

- North-south distances neither stretched nor compressed (1): equirectangular projection or "plate carrée".

- North-south compression equals the cosine of the latitude (the reciprocal of east-west stretching): equal-area cylindrical. This projection has many named specializations differing only in the scaling constant, such as the Gall–Peters or Gall orthographic (undistorted at the 45° parallels), Behrmann (undistorted at the 30° parallels), and Lambert cylindrical equal-area (undistorted at the equator). Since this projection scales north-south distances by the reciprocal of east-west stretching, it preserves area at the expense of shapes.

In the first case (Mercator), the east-west scale always equals the north-south scale. In the second case (central cylindrical), the north-south scale exceeds the east-west scale everywhere away from the equator. Each remaining case has a pair of secant lines—a pair of identical latitudes of opposite sign (or else the equator) at which the east-west scale matches the north-south-scale.

Normal cylindrical projections map the whole Earth as a finite rectangle, except in the first two cases, where the rectangle stretches infinitely tall while retaining constant width.

Pseudocylindrical

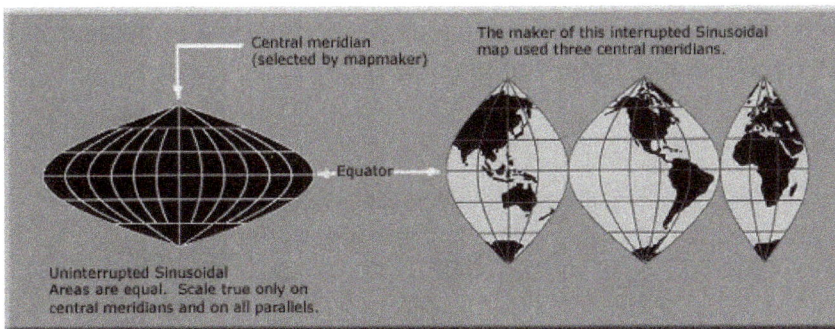

A sinusoidal projection shows relative sizes accurately, but grossly distorts shapes. Distortion can be reduced by "interrupting" the map.

Pseudocylindrical projections represent the *central* meridian as a straight line segment. Other meridians are longer than the central meridian and bow outward away from the central meridian. Pseudocylindrical projections map parallels as straight lines. Along parallels, each point from the surface is mapped at a distance from the central meridian that is proportional to its difference in longitude from the central meridian. Therefore, meridians are equally spaced along a given parallel. On a pseudocylindrical map, any point further from the equator than some other point has a higher latitude than the other point, preserving north-south relationships. This trait is useful when illustrating phenomena that depend on latitude, such as climate. Examples of pseudocylindrical projections include:

- Sinusoidal, which was the first pseudocylindrical projection developed. On the map, as in reality, the length of each parallel is proportional to the cosine of the latitude. The area of any region is true.

- Collignon projection, which in its most common forms represents each meridian as two straight line segments, one from each pole to the equator.

- Tobler
  hyperelliptical

- Mollweide

- Goode homolosine
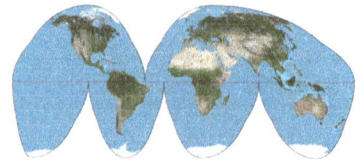

- Eckert IV
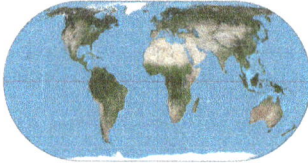

- Eckert VI

- Kavrayskiy VII

Hybrid

The HEALPix projection combines an equal-area cylindrical projection in equatorial regions with the Collignon projection in polar areas.

Conic

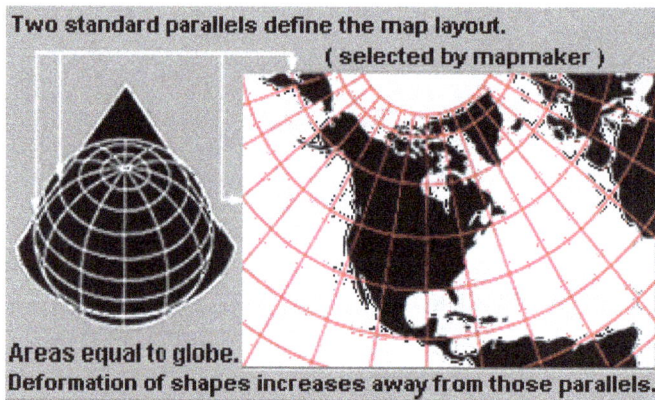

Albers conic.

The term "conic projection" is used to refer to any projection in which meridians are mapped to equally spaced lines radiating out from the apex and circles of latitude (parallels) are mapped to circular arcs centered on the apex.

When making a conic map, the map maker arbitrarily picks two standard parallels. Those standard parallels may be visualized as secant lines where the cone intersects the globe—or, if the map maker chooses the same parallel twice, as the tangent line where the cone is tangent to the globe. The resulting conic map has low distortion in scale, shape, and area near those standard parallels. Distances along the parallels to the north of both standard parallels or to the south of both standard parallels are stretched;

distances along parallels between the standard parallels are compressed. When a single standard parallel is used, distances along all other parallels are stretched.

Conic projections that are commonly used are:

- Equidistant conic, which keeps parallels evenly spaced along the meridians to preserve a constant distance scale along each meridian, typically the same or similar scale as along the standard parallels.

- Albers conic, which adjusts the north-south distance between non-standard parallels to compensate for the east-west stretching or compression, giving an equal-area map.

- Lambert conformal conic, which adjusts the north-south distance between non-standard parallels to equal the east-west stretching, giving a conformal map.

Pseudoconic

- Bonne

- Werner cordiform, upon which distances are correct from one pole, as well as along all parallels.

- American polyconic

Azimuthal (Projections onto a Plane)

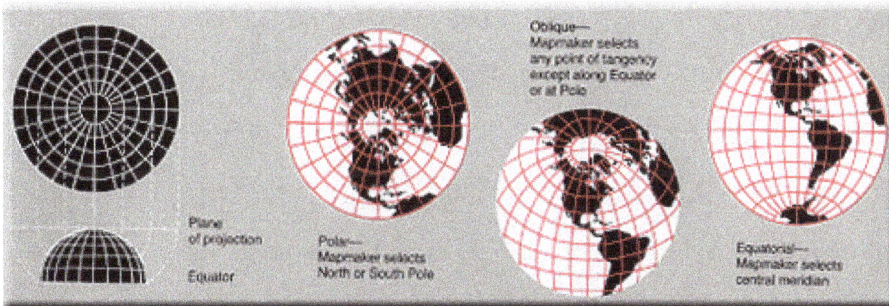

An azimuthal equidistant projection shows distances and directions accurately from the center point, but distorts shapes and sizes elsewhere.

Azimuthal projections have the property that directions from a central point are preserved and therefore great circles through the central point are represented by straight lines on the map. These projections also have radial symmetry in the scales and hence in the distortions: map distances from the central point are computed by a function $r(d)$ of the true distance d, independent of the angle; correspondingly, circles with the central point as center are mapped into circles which have as center the central point on the map.

The mapping of radial lines can be visualized by imagining a plane tangent to the Earth, with the central point as tangent point.

The radial scale is $r'(d)$ and the transverse scale $r(d) / \left(R \sin \dfrac{d}{R} \right)$ where R is the radius of the Earth.

Some azimuthal projections are true perspective projections; that is, they can be constructed mechanically, projecting the surface of the Earth by extending lines from a point of perspective (along an infinite line through the tangent point and the tangent point's antipode) onto the plane:

- The gnomonic projection displays great circles as straight lines. Can be constructed by using a point of perspective at the center of the Earth. $r(d) = c \tan \dfrac{d}{R}$; so that even just a hemisphere is already infinite in extent.

- The General Perspective projection can be constructed by using a point of perspective outside the earth. Photographs of Earth (such as those from the International Space Station) give this perspective.

- The orthographic projection maps each point on the earth to the closest point on the plane. Can be constructed from a point of perspective an infinite distance from the tangent point; $r(d) = C \sin \dfrac{d}{R}$. Can display up to a hemisphere on a finite circle. Photographs of Earth from far enough away, such as the Moon, approximate this perspective.

- The stereographic projection, which is conformal, can be constructed by using the tangent point's antipode as the point of perspective. $r(d) = c \tan \dfrac{d}{2R}$; the scale is $c / \left(2R \cos^2 \dfrac{d}{2R} \right)$. Can display nearly the entire sphere's surface on a finite circle. The sphere's full surface requires an infinite map.

Other azimuthal projections are not true perspective projections:

- Azimuthal equidistant: $r(d) = cd$; it is used by amateur radio operators to know the direction to point their antennas toward a point and see the distance to it. Distance from the tangent point on the map is proportional to surface distance on the earth (for the case where the tangent point is the North Pole)

- Lambert azimuthal equal-area. Distance from the tangent point on the map is proportional to straight-line distance through the earth: $r(d) = c \sin \dfrac{d}{2R}$

- Logarithmic azimuthal is constructed so that each point's distance from the center of the map is the logarithm of its distance from the tangent point on the

Earth. $r(d) = c \ln \dfrac{d}{d_0}$; locations closer than at a distance equal to the constant

d_0 a

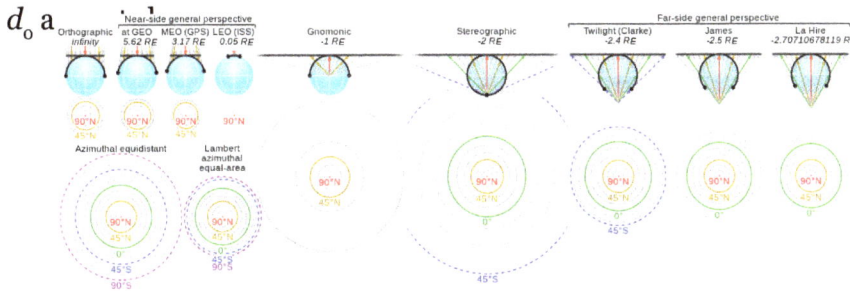

Comparison of some azimuthal projections centred on 90° N at the same scale, ordered by projection altitude in Earth radii.

Projections by Preservation of a Metric Property

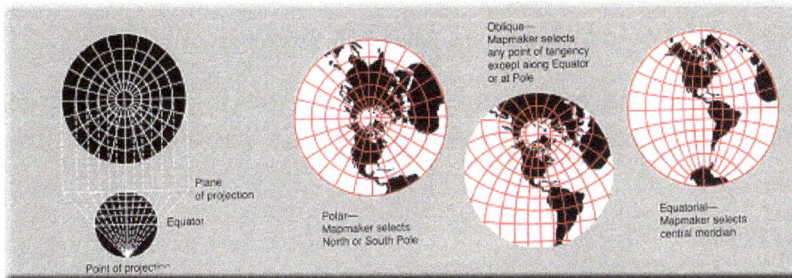

A stereographic projection is conformal and perspective but not equal area or equidistant.

Conformal

Conformal, or orthomorphic, map projections preserve angles locally, implying that they map infinitesimal circles of constant size anywhere on the Earth to infinitesimal circles of varying sizes on the map. In contrast, mappings that are not conformal distort most such small circles into ellipses of distortion. An important consequence of conformality is that relative angles at each point of the map are correct, and the local scale (although varying throughout the map) in every direction around any one point is constant. These are some conformal projections:

- Mercator: Rhumb lines are represented by straight segments

- Transverse Mercator

- Stereographic: Any circle of a sphere, great and small, maps to a circle or straight line.

- Roussilhe

- Lambert conformal conic

- Peirce quincuncial projection

- Adams hemisphere-in-a-square projection

- Guyou hemisphere-in-a-square projection

Equal-area

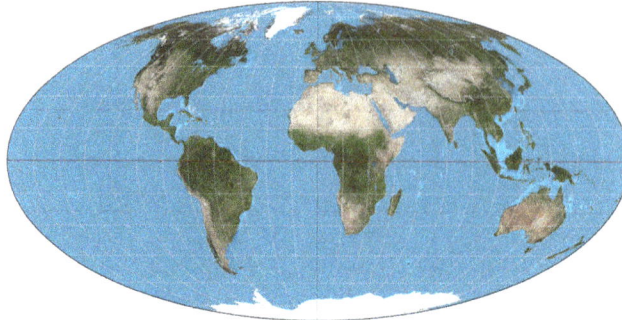

The equal-area Mollweide projection

Equal-area maps preserve area measure, generally distorting shapes in order to do that. Equal-area maps are also called *equivalent* or *authalic*. These are some projections that preserve area:

- Albers conic

- Bonne

- Bottomley

- Collignon

- Cylindrical equal-area

- Eckert II, IV and VI

- Gall orthographic (also known as Gall–Peters, or Peters, projection)

- Goode's homolosine

- Hammer

- Hobo–Dyer

- Lambert azimuthal equal-area

- Lambert cylindrical equal-area

- Mollweide

- Sinusoidal

- Snyder's equal-area polyhedral projection, used for geodesic grids.

- Tobler hyperelliptical

- Werner

Equidistant

A two-point equidistant projection of Eurasia

These are some projections that preserve distance from some standard point or line:

- Equirectangular—distances along meridians are conserved

- Plate carrée—an Equirectangular projection centered at the equator

- Azimuthal equidistant—distances along great circles radiating from centre are conserved

- Equidistant conic

- Sinusoidal—distances along parallels are conserved

- Werner cordiform distances from the North Pole are correct as are the curved distance on parallels

- Soldner

- Two-point equidistant: two "control points" are arbitrarily chosen by the map maker. Distance from any point on the map to each control point is proportional to surface distance on the earth.

Gnomonic

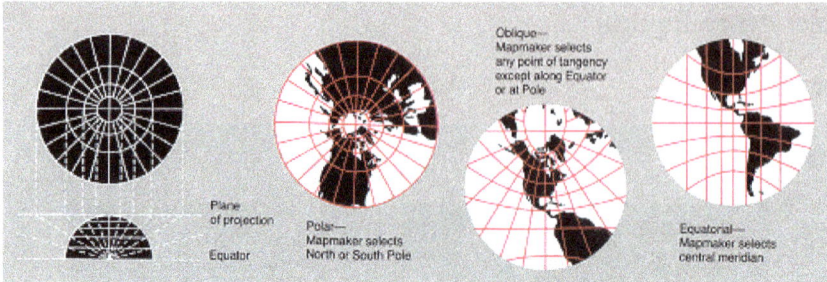

The Gnomonic projection is thought to be the oldest map projection, developed by Thales in the 6th century BC

Great circles are displayed as straight lines:

- Gnomonic projection

Retroazimuthal

Direction to a fixed location B (the bearing at the starting location A of the shortest route) corresponds to the direction on the map from A to B:

- Littrow—the only conformal retroazimuthal projection

- Hammer retroazimuthal—also preserves distance from the central point

- Craig retroazimuthal *aka* Mecca or Qibla—also has vertical meridians

Compromise Projections

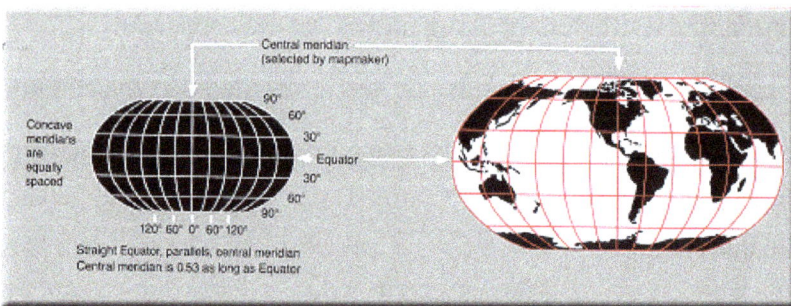

The Robinson projection was adopted by *National Geographic* magazine in 1988 but abandoned by them in about 1997 for the Winkel tripel.

Compromise projections give up the idea of perfectly preserving metric properties, seeking instead to strike a balance between distortions, or to simply make things "look right". Most of these types of projections distort shape in the polar regions more than at the equator. These are some compromise projections:

- Robinson

- van der Grinten

- Miller cylindrical

- Winkel Tripel

- Buckminster Fuller's Dymaxion

- B. J. S. Cahill's Butterfly Map

- Kavrayskiy VII projection

- Wagner VI projection

- Chamberlin trimetric

- Oronce Finé's cordiform

References

- "Maclure's geological map of the United States". US Library of Congress' Map Collection. Library of Congress. Retrieved 30 October 2015

- Bliss, Laura (13 October 2014). "These tiny glass globes were all the rage in London 200 years ago". Quartz (publication). Retrieved 2014-10-14

- Harrell, James A.; Brown, V. Max (1992). "The world's oldest surviving geological map—the 1150 BC Turin papyrus from Egypt". Journal of Geology. 100 (1): 3–18. JSTOR 30082315

- Flattening the Earth: Two Thousand Years of Map Projections, John P. Snyder, 1993, pp. 5-8, ISBN 0-226-76747-7. This is a survey of virtually all known projections from antiquity to 1993

- Government of Canada (2016-04-08). "National Topographic System Maps". Earth Sciences – Geography. Natural Resources Canada. Retrieved 2016-05-16

- Shingareva, K.B.; Bugaevsky, L.M.; Nyrtsov, M. (2000). "Mathematical Basis for Non-spherical Celestial Bodies Maps" (PDF). Journal of Geospatial Engineering. 2 (2): 45–50

- "The Four Great Surveys of the West". The United States Geological Survey: 1879-1989. U.S. Geological Survey, U.S. Department of the Interior. 2000-04-10. Retrieved 2007-06-19

- Snyder, John P. (1993). Flattening the earth: two thousand years of map projections. University of Chicago Press. ISBN 0-226-76746-9

- Hurst, Paul (2010), Will we be lost without paper maps in the digital age? (PDF) (M.S. thesis), U.K.: University of Sheffield, pp. 1–18, retrieved 2011-07-01

- Choosing a World Map. Falls Church, Virginia: American Congress on Surveying and Mapping. 1988. p. 1. ISBN 0-9613459-2-6

- Pickles, John. Cartography, Digital Transitions, and Questions of History (PDF). International Cartographic Association, 1999. Ottawa. p. 17. Retrieved 2011-06-29

Cartography: A Study of Maps

The science of making maps is known as cartography. With the advancement of technology such as the printing press and the Vernier scale, mass production of maps have become possible. Cartography can be divided into two categories, thematic cartography and general cartography. The chapter serves as a source to understand the major categories related to cartography.

Cartography

The above images shows a medieval depiction of the Ecumene (1482, Johannes Schnitzer, engraver), constructed after the coordinates in Ptolemy's Geography and using his second map projection. The translation into Latin and dissemination of *Geography* in Europe, in the beginning of the 15th century, marked the rebirth of scientific cartography, after more than a millennium of stagnation.

Cartography is the study and practice of making maps. Combining science, aesthetics, and technique, cartography builds on the premise that reality can be modeled in ways that communicate spatial information effectively.

The fundamental problems of traditional cartography are to:

- Set the map's agenda and select traits of the object to be mapped. This is the concern of map editing. Traits may be physical, such as roads or land masses, or may be abstract, such as toponyms or political boundaries.

- Represent the terrain of the mapped object on flat media. This is the concern of map projections.

- Eliminate characteristics of the mapped object that are not relevant to the map's purpose. This is the concern of generalization.

- Reduce the complexity of the characteristics that will be mapped. This is also the concern of generalization.

- Orchestrate the elements of the map to best convey its message to its audience. This is the concern of map design.

Modern cartography constitutes many theoretical and practical foundations of geographic information systems.

History

Copy (1472) of St. Isidore's TO map of the world.

The earliest known map is a matter of some debate, both because the term "map" isn't well-defined and because some artifacts that might be maps might actually be something else. A wall painting that might depict the ancient Anatolian city of Çatalhöyük (previously known as Catal Huyuk or Çatal Hüyük) has been dated to the late 7th millennium BCE. Among the prehistoric alpine rock carvings of Mount Bego (F) and Valcamonica (I), dated to the 4th millennium BCE, geometric patterns consisting of dotted rectangles and lines are widely interpreted in archaeological literature as a depiction of cultivated plots. Other known maps of the ancient world include the Minoan "House of the Admiral" wall painting from c. 1600 BCE, showing a seaside community in an oblique perspective and an engraved map of the holy Babylonian city of Nippur, from the Kassite period (14th – 12th centuries BCE). The oldest surviving world maps are from 9th century BCE Babylonia. One shows Babylon on the Euphrates, surrounded by Assyria, Urartu and several cities, all, in turn, surrounded by a "bitter river" (Oceanus). Another depicts Babylon as being north of the world center.

Valcamonica rock art (I), Paspardo r. 29, topographic composition, 4th millennium BCE

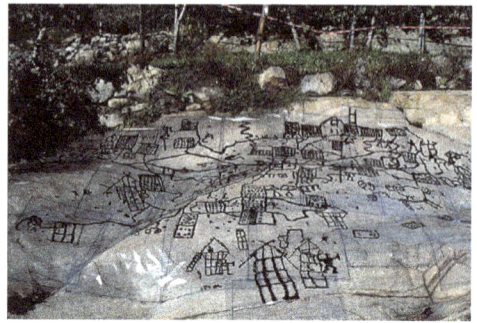

The *Bedolina Map* and its tracing, 6th–4th century BCE

The ancient Greeks and Romans created maps since Anaximander in the 6th century BCE. In the 2nd century AD, Ptolemy wrote his treatise on cartography, Geographia. This contained Ptolemy's world map – the world then known to Western society *(Ecumene)*. As early as the 8th century, Arab scholars were translating the works of the Greek geographers into Arabic.

In ancient China, geographical literature dates to the 5th century BCE. The oldest extant Chinese maps come from the State of Qin, dated back to the 4th century BCE, during the Warring States period. In the book of the *Xin Yi Xiang Fa Yao*, by the Chinese scientist Su Song, a star map on the equidistant cylindrical projection. Although this method of charting seems to have existed in China even prior to this publication and scientist, the greatest significance of the star maps by Su Song is that they represent the oldest existent star maps in printed form.

Early forms of cartography of India included depictions of the pole star and surrounding constellations. These charts may have been used for navigation.

Mappa mundi are the Medieval European maps of the world. Approximately 1,100 mappae mundi are known to have survived from the Middle Ages. Of these, some 900 are found illustrating manuscripts and the remainder exist as stand-alone documents.

The *Tabula Rogeriana*, drawn by Muhammad al-Idrisi for Roger II of Sicily in 1154

The Arab geographer Muhammad al-Idrisi produced his medieval atlas *Tabula Rogeriana* in 1154. He incorporated the knowledge of Africa, the Indian Ocean and the Far

East, gathered by Arab merchants and explorers with the information inherited from the classical geographers to create the most accurate map of the world up until his time. It remained the most accurate world map for the next three centuries.

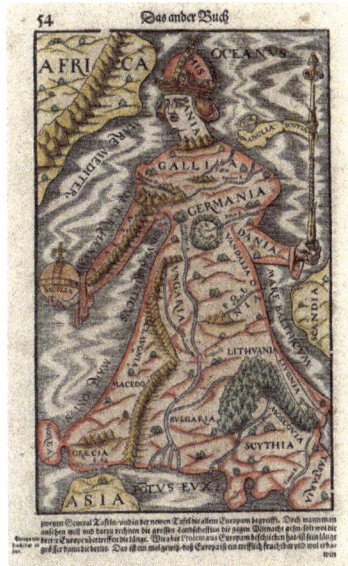

Europa regina in Sebastian Münster's "*Cosmographia*", 1570

In the Age of Exploration, from the 15th century to the 17th century, European cartographers both copied earlier maps (some of which had been passed down for centuries) and drew their own based on explorers' observations and new surveying techniques. The invention of the magnetic compass, telescope and sextant enabled increasing accuracy. In 1492, Martin Behaim, a German cartographer, made the oldest extant globe of the Earth.

Johannes Werner refined and promoted the Werner projection. In 1507, Martin Waldseemüller produced a globular world map and a large 12-panel world wall map (*Universalis Cosmographia*) bearing the first use of the name "America". Portuguese cartographer Diego Ribero was the author of the first known planisphere with a graduated Equator (1527). Italian cartographer Battista Agnese produced at least 71 manuscript atlases of sea charts.

Due to the sheer physical difficulties inherent in cartography, map-makers frequently lifted material from earlier works without giving credit to the original cartographer. For example, one of the most famous early maps of North America is unofficially known as the "Beaver Map", published in 1715 by Herman Moll. This map is an exact reproduction of a 1698 work by Nicolas de Fer. De Fer in turn had copied images that were first printed in books by Louis Hennepin, published in 1697, and François Du Creux, in 1664. By the 18th century, map-makers started to give credit to the original engraver by printing the phrase "After [the original cartographer]" on the work.

Technological Changes

The image above shows a pre-Mercator nautical chart of 1571, from Portuguese cartographer Fernão Vaz Dourado (c. 1520–c. 1580). It belongs to the so-called *plane chart* model, where observed latitudes and magnetic directions are plotted directly into the plane, with a constant scale, as if the Earth were a plane (Portuguese National Archives of Torre do Tombo, Lisbon).

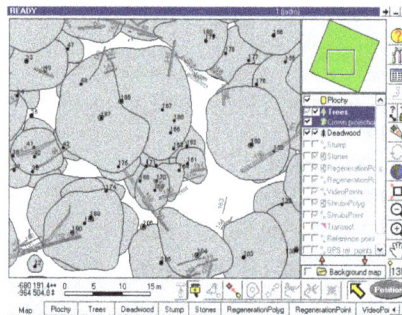

Mapping can be done with GPS and laser rangefinder directly in the field. Image shows mapping of forest structure (position of trees, dead wood and canopy).

In cartography, technology has continually changed in order to meet the demands of new generations of mapmakers and map users. The first maps were manually constructed with brushes and parchment; therefore, varied in quality and were limited in distribution. The advent of magnetic devices, such as the compass and much later, magnetic storage devices, allowed for the creation of far more accurate maps and the ability to store and manipulate them digitally.

Advances in mechanical devices such as the printing press, quadrant and vernier, allowed for the mass production of maps and the ability to make accurate reproductions from more accurate data. Optical technology, such as the telescope, sextant and other devices that use telescopes, allowed for accurate surveying of land and the ability of mapmakers and navigators to find their latitude by measuring angles to the North Star at night or the sun at noon.

Advances in photochemical technology, such as the lithographic and photochemical processes, have allowed for the creation of maps that have fine details, do not distort in shape and resist moisture and wear. This also eliminated the need for engraving, which further shortened the time it takes to make and reproduce maps.

In the 20th century, aerial photography, satellite imagery, and remote sensing provided efficient, precise methods for mapping physical features, such as coastlines, roads, buildings, watersheds, and topography. Advancements in electronic technology ushered in another revolution in cartography. Ready availability of computers and peripherals such as monitors, plotters, printers, scanners (remote and document) and analytic stereo plotters, along with computer programs for visualization, image processing, spatial analysis, and database management, democratized and greatly expanded the making of maps. The ability to superimpose spatially located variables onto existing maps created new uses for maps and new industries to explore and exploit these potentials.

These days most commercial-quality maps are made using software that falls into one of three main types: CAD, GIS and specialized illustration software. Spatial information can be stored in a database, from which it can be extracted on demand. These tools lead to increasingly dynamic, interactive maps that can be manipulated digitally.

With the field rugged computers, GPS and laser rangefinders, it is possible to perform mapping directly in the terrain.

Deconstruction

There are technical and cultural aspects to the producing maps. In this sense, maps are biased. The study of bias, influence, and agenda in making a map is what comprise a map's deconstruction. A central tenet of deconstructionism is that maps have power. Other assertions are that maps are inherently biased and that we search for metaphor and rhetoric in maps.

It was the Europeans who promoted an epistemological understanding of the map as early as the 17th century. An example of this understanding is that, "[European reproduction of terrain on maps] reality can be expressed in mathematical terms; that systematic observation and measurement offer the only route to cartographic truth...". 17th century map-makers were careful and precise in their strategic approaches to maps based on a scientific model of knowledge. Popular belief at the time was that this scientific approach to cartography was immune to the social atmosphere.

A common belief is that science heads in a direction of progress, and thus leads to more accurate representations of maps. In this belief European maps must be superior to others, which necessarily employed different map-making skills. "There was a 'not cartography' land where lurked an army of inaccurate, heretical, subjective, valuative, and ideologically distorted images. Cartographers developed a 'sense of the other' in relation to nonconforming maps."

Though cartography has been a target of much criticism in recent decades, a cartographer's 'black box' always seemed to be naturally defended to the point where it overcame the criticism. However, to later scholars in the field, it was evident that cultural influences dominate map-making. For instance, certain abstracts on maps and the map-making society itself describe the social influences on the production of maps. This social play on cartographic knowledge "...produces the 'order' of [maps'] features and the 'hierarchies of its practices.'"

Depictions of Africa are a common target of deconstructionism. According to deconstructionist models, cartography was used for strategic purposes associated with imperialism and as instruments and representations of power during the conquest of Africa. The depiction of Africa and the low latitudes in general on the Mercator projection has been interpreted as imperialistic and as symbolic of subjugation due to the diminished proportions of those regions compared to higher latitudes where the European powers were concentrated.

Maps furthered imperialism and colonization of Africa through practical ways such as showing basic information like roads, terrain, natural resources, settlements, and communities. Through this, maps made European commerce in Africa possible by showing potential commercial routes, and made natural resource extraction possible by depicting locations of resources. Such maps also enabled military conquests and made them more efficient, and imperial nations further used them to put their conquests on display. These same maps were then used to cement territorial claims, such as at the Berlin Conference of 1884–1885.

Before 1749, maps of the African continent had African kingdoms drawn with assumed or contrived boundaries, with unknown or unexplored areas having drawings of animals, imaginary physical geographic features, and descriptive texts. In 1748 Jean B. B. d'Anville created the first map of the African continent that had blank spaces to represent the unknown territory. This was revolutionary in cartography and the representation of power associated with map making.

Map Types

General vs. Thematic Cartography

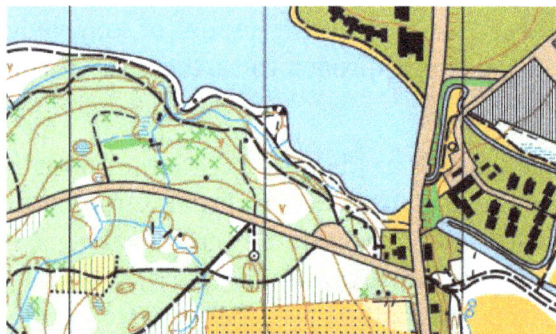

Small section of an orienteering map.

In understanding basic maps, the field of cartography can be divided into two general categories: general cartography and thematic cartography. General cartography involves those maps that are constructed for a general audience and thus contain a variety of features. General maps exhibit many reference and location systems and often are produced in a series. For example, the 1:24,000 scale topographic maps of the United States Geological Survey (USGS) are a standard as compared to the 1:50,000 scale Canadian maps. The government of the UK produces the classic 1:50,000 (replacing the older 1 inch to 1 mile) "Ordnance Survey" maps of the entire UK and with a range of correlated larger- and smaller-scale maps of great detail. Many private mapping companies have also produce thematic map series.

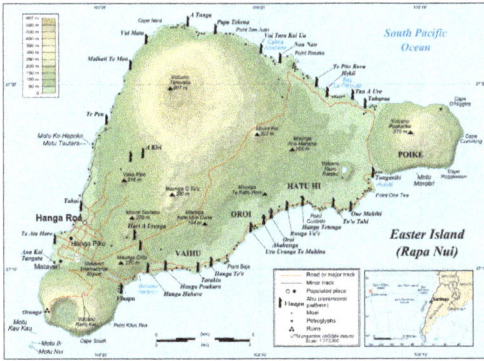

Topographic map of Easter Island · · · · · · · · · · · Relief map Sierra Nevada

Thematic cartography involves maps of specific geographic themes, oriented toward specific audiences. A couple of examples might be a dot map showing corn production in Indiana or a shaded area map of Ohio counties, divided into numerical choropleth classes. As the volume of geographic data has exploded over the last century, thematic cartography has become increasingly useful and necessary to interpret spatial, cultural and social data.

An orienteering map combines both general and thematic cartography, designed for a very specific user community. The most prominent thematic element is shading, that indicates degrees of difficulty of travel due to vegetation. The vegetation itself is not identified, merely classified by the difficulty ("fight") that it presents.

Topographic vs. Topological

A topographic map is primarily concerned with the topographic description of a place, including (especially in the 20th and 21st centuries) the use of contour lines showing elevation. Terrain or relief can be shown in a variety of ways.

A topological map is a very general type of map, the kind one might sketch on a napkin. It often disregards scale and detail in the interest of clarity of communicating specific route or relational information. Beck's London Underground map is an iconic example. Though the most widely used map of "The Tube," it preserves little of reality: it varies scale constantly and abruptly, it straightens curved tracks, and it contorts directions. The only topography on it is the River Thames, letting the reader know whether a station is north or south of the river. That and the topology of station order and interchanges between train lines are all that is left of the geographic space. Yet those are all a typical passenger wishes to know, so the map fulfils its purpose.

Map Design

Illustrated map.

Map Purpose and Selection of Information

Arthur H. Robinson, an American cartographer influential in thematic cartography, stated that a map not properly designed "will be a cartographic failure." He also claimed, when considering all aspects of cartography, that "map design is perhaps the most complex." Robinson codified the mapmaker's understanding that a map must be designed foremost with consideration to the audience and its needs.

From the very beginning of mapmaking, maps "have been made for some particular purpose or set of purposes". The intent of the map should be illustrated in a manner in which the percipient acknowledges its purpose in a timely fashion. The term *percipient* refers to the person receiving information and was coined by Robinson. The principle of figure-ground refers to this notion of engaging the user by presenting a clear presentation, leaving no confusion concerning the purpose of the map. This will enhance the user's experience and keep his attention. If the user is unable to identify what is being demonstrated in a reasonable fashion, the map may be regarded as useless.

Making a meaningful map is the ultimate goal. Alan MacEachren explains that a well designed map "is convincing because it implies authenticity" (1994, pp. 9). An interesting map will no doubt engage a reader. Information richness or a map that is multivariate shows relationships within the map. Showing several variables allows comparison, which adds to the meaningfulness of the map. This also generates hypothesis and stimulates ideas and perhaps further research. In order to convey the message of the map, the creator must design it in a manner which will aid the reader in the overall understanding of its purpose. The title of a map may provide the "needed link" necessary for communicating that message, but the overall design of the map fosters the manner in which the reader interprets it (Monmonier, 1993, pp. 93).

In the 21st century it is possible to find a map of virtually anything from the inner workings of the human body to the virtual worlds of cyberspace. Therefore, there are now a huge variety of different styles and types of map – for example, one area which has evolved a specific and recognisable variation are those used by public transport organisations to guide passengers, namely urban rail and metro maps, many of which are loosely based on 45 degree angles as originally perfected by Harry Beck and George Dow.

Naming Conventions

Most maps use text to label places and for such things as the map title, legend and other information. Although maps are often made in one specific language, place names often differ between languages. So a map made in English may use the name *Germany* for that country, while a German map would use *Deutschland* and a French map *Allemagne*. A non-native term for a place is referred to as an exonym.

In some cases the correct name is not clear. For example, the nation of Burma officially changed its name to Myanmar, but many nations do not recognize the ruling junta and continue to use *Burma*. Sometimes an official name change is resisted in other languages and the older name may remain in common use. Examples include the use of *Saigon* for Ho Chi Minh City, *Bangkok* for Krung Thep and *Ivory Coast* for Côte d'Ivoire.

Difficulties arise when transliteration or transcription between writing systems is required. Some well-known places have well-established names in other languages and writing systems, such as *Russia* or *Rußland* for Росси́я, but in other cases a system of transliteration or transcription is required. Even in the former case, the exclusive use of an exonym may be unhelpful for the map user. It will not be much use for an English user of a map of Italy to show Livorno *only* as "Leghorn" when road signs and railway timetables show it as "Livorno". In transliteration, the characters in one script are represented by characters in another. For example, the Cyrillic letter *P* is usually written as *R* in the Latin script, although in many cases it is not as simple as a one-for-one equivalence. Systems exist for transliteration of Arabic, but the results may vary. For example, the Yemeni city of Mocha is written variously in English as Mocha, Al Mukha, al-Mukhā, Mocca and Moka. Transliteration systems are based on relating written symbols to one another, while transcription is the attempt to spell in one language the phonetic sounds of another. Chinese writing is now usually converted to the Latin alphabet through the Pinyin phonetic transcription systems. Other systems were used in the past, such as Wade-Giles, resulting in the city being spelled *Beijing* on newer English maps and *Peking* on older ones.

Further difficulties arise when countries, especially former colonies, do not have a strong national geographic naming standard. In such cases, cartographers may have to choose between various phonetic spellings of local names versus older imposed, sometimes resented, colonial names. Some countries have multiple official languages, resulting in multiple official placenames. For example, the capital of Belgium is both *Brussel* and *Bruxelles*. In Canada, English and French are official languages and places have names in both languages. British Columbia is also officially named *la Colombie-Britannique*. English maps rarely show the French names outside of Quebec, which itself is spelled *Québec* in French.

The study of placenames is called toponymy, while that of the origin and historical usage of placenames as words is etymology.

In order to improve legibility or to aid the illiterate, some maps have been produced using pictograms to represent places. The iconic example of this practice is Lance Wyman's early plans for the Mexico City Metro, on which stations were shown simply as stylized logos. Wyman also prototyped such a map for the Washington Metro, though ultimately the idea was rejected. Other cities experimenting with such maps are Fukuoka, Guadalajara and Monterrey.

Map Symbology

Cartographic symbology encodes information on the map in ways intended to convey information to the map reader efficiently, taking into consideration the limited space on the map, models of human understanding through visual means, and the likely cultural background and education of the map reader. Symbology may be implicit, using universal elements of design, or may be more specific to cartography or even to the map.

A map may have any of many kinds of symbolization. Some examples are:

- A legend, or key, explains the map's pictorial language.

- A title indicates the region and perhaps the theme that the map portrays.

- A neatline frames the entire map image.

- A compass rose or north arrow provides orientation.

- An overview map gives global context for the primary map.

- A bar scale translates between map measurements and real distances.

- A map projection provides a way to represent the curved surface on the plane of the map.

The map may declare its sources, accuracy, publication date and authorship, and so forth. The map image itself portrays the region.

Map coloring is another form of symbology, one whose importance can reach beyond aesthetic. In complex thematic maps, for example, the color scheme's structure can critically affect the reader's ability to understand the map's information. Modern computer displays and print technologies can reproduce much of the gamut that humans can perceive, allowing for intricate exploitation of human visual discrimination in order to convey detailed information.

Quantitative symbols give a visual indication of the magnitude of the phenomenon that the symbol represents. Two major classes of symbols are used to portray quantity. Proportional symbols change size according to phenomenon's magnitude, making them appropriate for representing statistics. Choropleth maps portray data collection areas, such as counties or census tracts, with color. Using color this way, the darkness and intensity (or value) of the color is evaluated by the eye as a measure of intensity or concentration.

Map Key or Legend

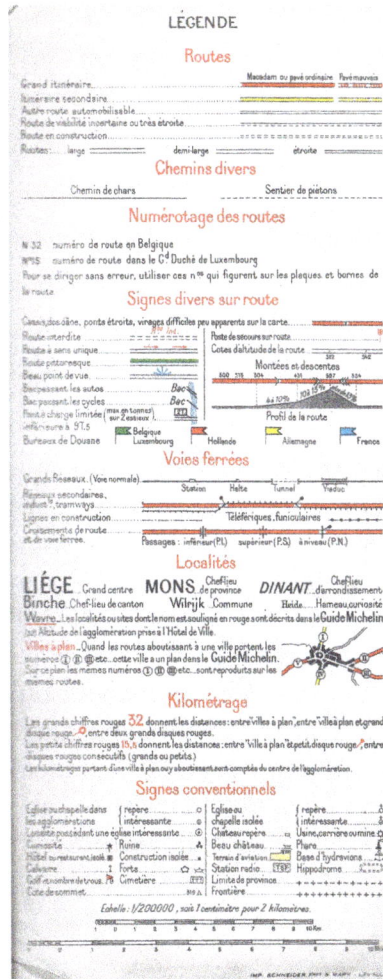

Legend or key of a French road map (Michelin 1940)

The map key, or legend, describes how to interpret the map's symbols and may give details of publication and authorship.

Examples of Point Symbols

Symbol	Explanation
⚒⚒	mine (Hammer and pick symbol), former mine
♟ ♙	castle, Burg
♱ ♱	church, chapel, monastery (♁)
⚇	monument
⌂	Hotel
✈✈	airport
🚂🚂	railway station
ⓘⓘ	Tourist information

Map Generalization

A good map has to compromise between portraying the items of interest (or themes) in the right place on the map, and the need to show that item using text or a symbol, which take up space on the map and might displace some other item of information. The cartographer is thus constantly making judgements about what to include, what to leave out and what to show in a *slightly* incorrect place. This issue assumes more importance as the scale of the map gets smaller (i.e. the map shows a larger area) because the information shown on the map takes up more space *on the ground*. A good example from the late 1980s was the Ordnance Survey's first digital maps, where the *absolute* positions of major roads were sometimes a scale distance of hundreds of metres away from ground truth, when shown on digital maps at scales of 1:250,000 and 1:625,000, because of the overriding need to annotate the features.

Map Projections

The Earth being spherical, any flat representation generates distortions such that shapes and areas cannot both be conserved simultaneously, and distances can never all be preserved. The mapmaker must choose a suitable *map projection* according to the space to be mapped and the purpose of the map.

Cartographic Errors

Some maps contain deliberate errors or distortions, either as propaganda or as a "watermark" to help the copyright owner identify infringement if the error appears in competitors' maps. The latter often come in the form of nonexistent, misnamed, or misspelled "trap streets". Other names and forms for this are paper townsites, fictitious entries, and copyright easter eggs.

Another motive for deliberate errors is cartographic "vandalism": a mapmaker wishing to leave his or her mark on the work. Mount Richard, for example, was a fictitious peak on the Rocky Mountains' continental divide that appeared on a Boulder County, Colorado map in the early 1970s. It is believed to be the work of draftsman Richard Ciacci. The fiction was not discovered until two years later.

Sandy Island (New Caledonia) is an example of a fictitious location that stubbornly survives, reappearing on new maps copied from older maps while being deleted from other new editions.

Cartography is the art and science of map making. Cartographers make a huge contribution in making the maps more meaningful and understandable. In this major principles of cartography are introduced.

Data Classification

It is important to have a good understanding of the data which needs to be represented on a map. One must recognize the scale of measurement for a particular data set because the scales determine the kind of mathematical operations that can be performed on the data. These scales of measurement are described below:

Nominal Scale

It only satisfies the identity property. The values assigned to variables are descriptive; they cannot be used for mathematical comparisons. For example, on the basis of gender we can classify individuals into 'male' and 'female' but neither of the gender is higher or smaller than the other.

Ordinal Scale

This scale has the properties of both identity and magnitude but the interval between any two values is indeterminate. It generally uses the operators "greater than", "equal to" or "less than" for ordering the observations. For example, the result of an athletic event declares the ranks secured by various athletes but the ranks themselves do not describe by what time a rank holder has finished the race with respect to the another athlete.

Interval Scale

This scale has the properties of identity, magnitude and equal intervals. The Fahrenheit scale for measuring temperature is made up of equal temperature units, so the difference between 10 and 20 degrees Fahrenheit is equal to the difference between 40 and 50 degrees Fahrenheit.

The use of interval scale tells whether one is greater or smaller than the other and it also quantifies the amount by which one is greater or smaller than the other. For example the temperature of a city recorded on the first day of a month is 15 degrees and on the second day is 20 degrees. So we can say that the temperature on the second day is higher than that on the first and the second day is 5 degrees hotter than the first day.

Ratio Scale

The ratio scale has the properties of identity, magnitude, equal interval and absolute zero. Having true zero allows for computing ratios. Weight is an example of ratio scale. Weights can be given ranks, units along the weight scale are equal to each other and weight has an absolute zero.

0 kg means there is no weight and 100 kg is said to be twice of 50 kg.

Map Layout

A map conveys geographical information and relationships. The result of any analysis in GIS is communicated using maps so as to help users/readers to better understand the geographical phenomenon. The map can only fulfill its purpose when it is presented in a proper manner. There are certain points which are to be kept in mind while designing a map:

a. Objective: It is necessary to pay attention to the question why and where will the map going to be used. One should be clear about whether the map is just sharing information, depicting result of an analysis or highlighting the key issues and relationships. One should also know whether the map is going to be displayed on a wall, in a book or in some other information resource.

b. Audience: A designer/cartographer must know who will be addressed through the maps. Maps must match the level/expectations of the audience. The level of design for a technical group and for general people ought to be different.

c. Balanced design: A map must be prepared using appropriate page size, color, patterns of shading, text, and scale. Everything on the map should be legible.

Designing the Map Layout

The process of map composition starts with preparing a layout for the map. Apart from the data, a map has certain other things that make map a package of effective and clear communication. These provide critical information to users and are known as map elements. A layout specifies the space and positions for different map elements such as neat lines, title, North arrow, scale bar, legend etc. Preparing an effective layout often requires experimentation with the available space.

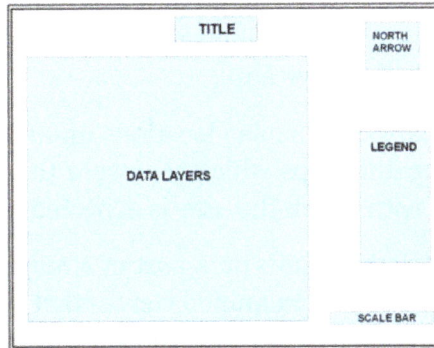

A map Layout

Every element in the layout has to be given a thought—whether it is important to include the respective element, does it require elaboration etc. The map elements that are generally found on a map are:

- Scale: Scale must be given in order to derive the actual size of an entity on the map or distance between two geographical entities on a map.

- Direction: True north is the direction of North Pole and it differs from the magnetic north. The magnetic north pole changes due to the changes in the geo-physical condition of the earth. Many maps indicate both the true and the magnetic north but the direction that is indicated on most of the maps is the true north.

- Legend: Legend lists all the symbols used in a map and describes what they depict.

- Title: A short suitable text that clearly defines the theme of the map.

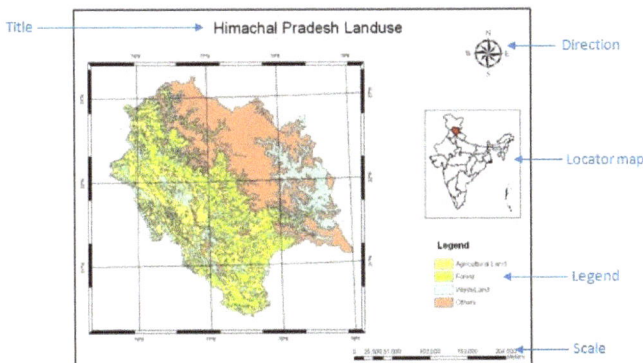

A map with its elements

Given below are a few other elements that are selectively used:

- Neatlines: These are lines used to frame a map to indicate where the map begins and where it ends.

- Reference grid/Graticule: A reference grid is a network of evenly spaced horizontal and vertical lines used to create context on a map. The grid can be used to show unique locations (control points) either in a geographic coordinate system or in a projected coordinate system.

- Locator map: The maps that display locations unfamiliar to users, are generally supported with locator maps which represent the locations in a larger geographic context and with which the user is expected to be familiar.

- Inset map: Sometimes the details on a part of a map are so clustered that they become difficult to read and a magnified view of that part of the map is required. These magnified view maps or close up maps are called insets.

- Source of information: The age, accuracy and reliability of the data sources are critical in carrying out any study. One can show the sources of data one has used in maps so that a user can track them and check his analysis and interpretation.

- Date of production: The representation of time on the maps is important in some cases. For example the weather map prepared on daily, weekly or monthly basis must indicate time on them. A road map to be used for a developing city must be a recent one so as to be relevant for the city etc.

Map Elements

Color

The aim of filling colors in a map is to make visual distinction among various features thus making map more decipherable.

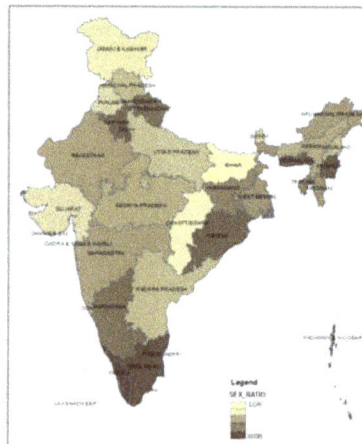

Map showing sex ratio in different states of India

Hue is the dominant wavelength we usually call as color such as green, red or blue. Saturation is the purity of hue or the dominance of hue. Value measures how dark or light the color is when hue is maintained constant. Changes in hue usually indicate qualitative differences such as different administrative units whereas changes in value and saturation represent quantitative differences such as population density in a country.

The lightness or darkness of the color represents quantitative differences. Generally, dark colors mean more and represent high values of the attribute under study. The above given map depicts the sex ratio in various states of India. The areas in dark color represent states with high sex ratio and the ones in light color represent states with low sex ratio.

There is a slight difference in color that appears when a map is displayed on a computer screen and when it is printed on a paper. The difference is attributed to the two different ways of creating colors known as additive, and subtractive models of color which are explained below.

Additive Color Model

It involves light emitted from a source and is employed on devices that use light such as camera, monitors etc. The three primary colors red, green, and blue are mixed at different levels to generate other colors. When one of these colors is combined with the other in equal amounts the secondary colors cyan, magenta, and yellow are produced. The combination of all the three colors in equal intensities produces white.

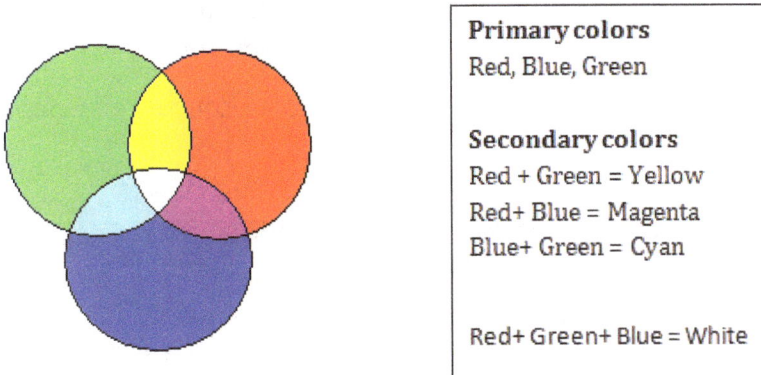

Primary colors
Red, Blue, Green

Secondary colors
Red + Green = Yellow
Red+ Blue = Magenta
Blue+ Green = Cyan

Red+ Green+ Blue = White

Additive color mixing

Subtractive Color Model

A subtractive color model explains the mixing of dyes, inks etc. to create a range of colors, each formed by subtracting i.e. absorbing some wavelengths of light and reflecting the others. A printer works on this color model. It uses three colors—cyan, magenta and yellow. When these are mixed together black is seem to be formed but the color is actually dark brown. This is why most printers add a black ink to print black text.

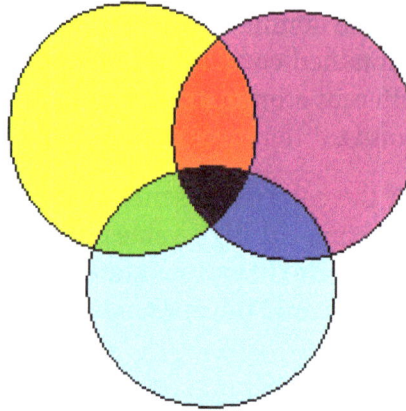

Subtractive color mixing

Text

The use of text on maps enhances the information one wishes to convey. Displaying geographic features and symbols alone on the map doesn't convey the full meaning until and unless it is supported with the relevant text. Inserting text on maps can seve various purposes such as:

- To label the features on map layers, e.g. labeling the name of the states in the political map of India

- To use text as graphic so as to highlight a particular area, e.g. labeling the location of Taj Mahal

- To add information such as title, author, data source references to the map layout.

Placement of Labels

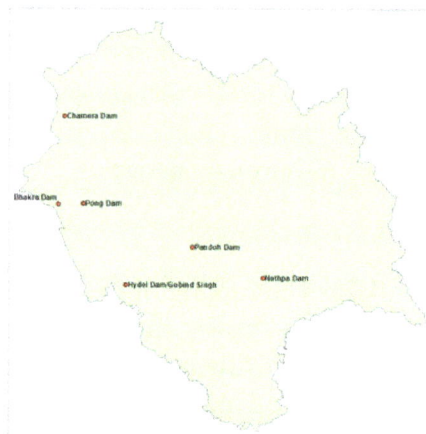

Proper placement of labels that mark the dams on Himachal Pradesh map

Placement of labels using call outs that mark lakes in Himachal Pradesh map

Labeling refers to placing of a descriptive text onto a feature on the map. The placement of labels is an important part of cartographic design because labeling affects the readability of a map. Placing labels at appropriate positions facilitates users to associate labels with the features being described. Sometimes, a part of map becomes overcrowded with labels making it difficult for a user to distinguish which label is used for which feature. This situation can be taken care of in a dynamic map, by adjusting labels in such a manner that they only appear on the map after it is viewed on a particular scale. For static maps, call outs are used for labeling features in such a situation.

Symbols

A symbol is a graphic or a pictorial element used to represent a feature on map. Various types of symbols are used for representing objects or features belonging to any of the three themes viz. point, line and polygon. Every symbol has a set of properties associated with it. These include its shape, size, color, angle, pattern etc.

Shape is the geometric form of the symbol. It is used to differentiate between the object classes. The closer the shapes of the symbols resemble the features they represent, the better is the map perceived by the users.

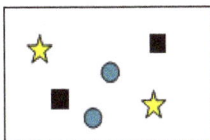

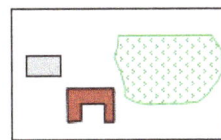

Point symbols   Line symbols   Polygon symbols

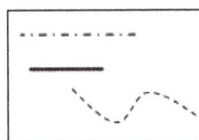

Symbols for different features

The size of symbols depicts a quantitative difference in the distribution of an attribute. Given below is a map showing petrol pump sites in an area. The difference in size of symbols that represent petrol pump stations corresponds to the number of people working in the particular petrol pump station. The biggest sized symbol represents that the station has the highest manpower as compared to the other stations.

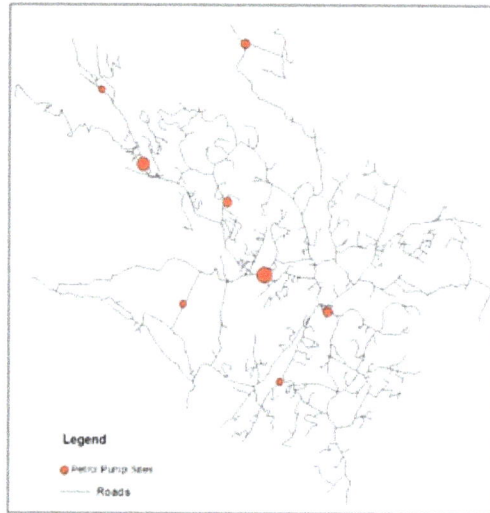

Attribute (manpower) representation by size of symbol

In the following map bars have been used to show the comparison and distribution of male and female populations in different states of India. The size of a bar corresponds to the number of individuals (male or female, whatever the bar represents) existing in a particular state.

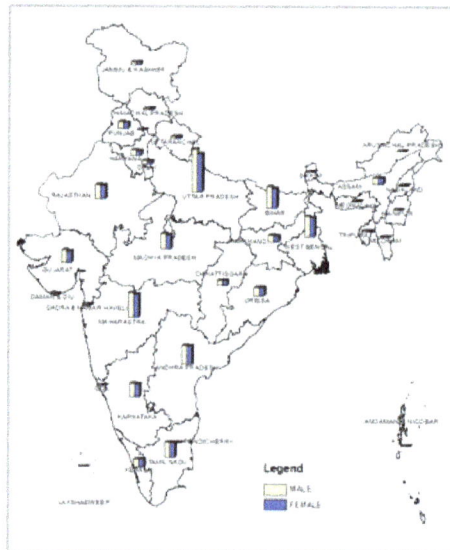

Attribute (male/female population) representation by bar size

Sometimes a combination of symbol and color exhibit the attribute of interest. In the example shown below, the elevation gradation of a landscape is marked by the simultaneous use of contours (lines) and colors.

Elevation gradation of a landscape

Data Presentation

After the careful preparation on several mapping tools which are integrated with GIS, the maps are presented to users. The final maps are of high cartographic quality and are brought out using a wide range of devices. Some of these devices are as follows:

Visual Display Unit

The results of mapping and GIS analysis may be presented on the computers over the internet for people who work online. Computer screens use cathode ray tube (CRT) technology or the Light emitting diodes (LEDs) to form images on the screen. The difference in output resulting from these devices depends on the hardware and GIS display software used by the computers. The hardware and software determine resolution of the screen which controls the detail that may be displayed, the number of colors used and size and scale of the data shown.

National Map Viewer is a platform for visualization of country maps on the computers over the internet. It is a GIS application designed by USGS which provides easy access to all the public domain geospatial information about United States. The National Map Viewer is a web portal that can be accessed online for geospatial information which also allows download of digital data and creation of cartographic products. Such application disseminates accurate and critical information to end users.

Plotters

Plotters are output devices for making copies of geographical data on paper or film. Plotters hold paper either through a roller or a flat bed surface. The drawing arm has colored pens. The two dimensional line images are created as per the commands given by the software. Few plotters contain preprogrammed information for complex shapes and symbol drawing which need reference by a computer command. This makes a plotter fast and flexible. Plotters now have bubble jet drawing devices in place of colored pens which allow faster and uniform drawing.

Plotter

Printers

A printer is an output device that prints an electronically stored document on print media such as paper or transparencies. The toner based printer or the laser printer take the output from the computer processor and convert it into the laser signal which is imprinted using a scanning action onto an electrically charged drum. In case of inkjet printers with response to the electrical signals from the GIS software the colored inks are emitted from the nozzle in the print head and the inks are transferred from the cartridge to the paper.

Printer

Toposheet Indexing

Survey of India produces the topographic maps of India. These maps are produced at different scales. In order to identify a map of a particular area, a numbering system has been adopted by the Survey of India.

For the purpose of an international series (within 4° N to 40° N Latitude and 44° E to 124° E Longitude) at the scale of 1: 1,000,000 is considered as a base map. This map is divided into sections of 4° latitude × 4° longitude and designated from 1 to 136 consisting of the segments that cover only land area.

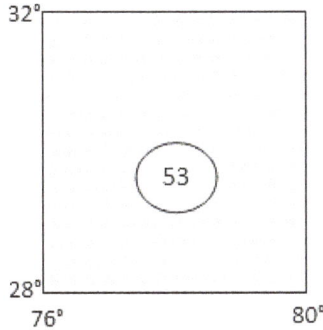

A sheet of 4° × 4° (scale: 1: 1,000,000)

Each section is further divided into 16 sections (4 rows and 4 columns) each of 1° latitude × 1° longitude. The sections start from Northwest direction, run column wise and end in Southeast direction.

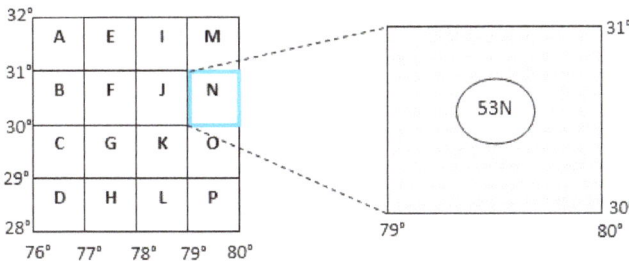

A sheet of 4° × 4° (scale: 1: 1,000,000)   A sheet of 1° × 1° (scale: 1: 250, 000)

The 1°×1° sheets are further subdivided into four parts, each of 30′ latitude × 30′ longitude. These are identified by the cardinal directions NE, NW, SE and SW.

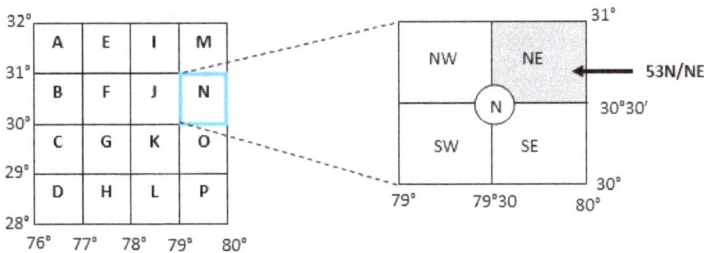

A sheet of 4° × 4° (scale: 1: 1,000,000)   A sheet of 30′ × 30′ (scale: 1:100,000)

The 1°×1° sheets can also be divided into 16 sections each of 15′ latitude × 15′ longitude and are numbered from 1 to 16 in a columned manner.

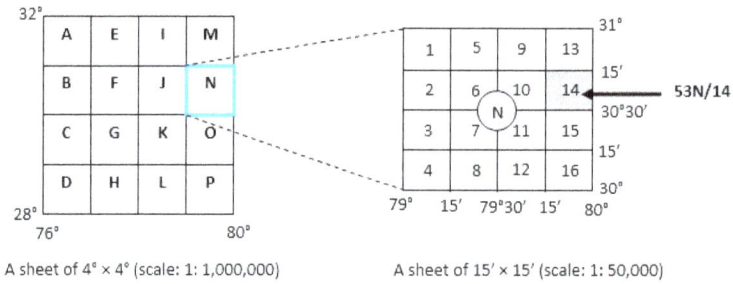

A sheet of 4° × 4° (scale: 1: 1,000,000)   A sheet of 15' × 15' (scale: 1: 50,000)

A 15'×15' sheet can be divided into 4 sheets, each of 7(1/2)' and are numbered as NW, NE, SW and SE.

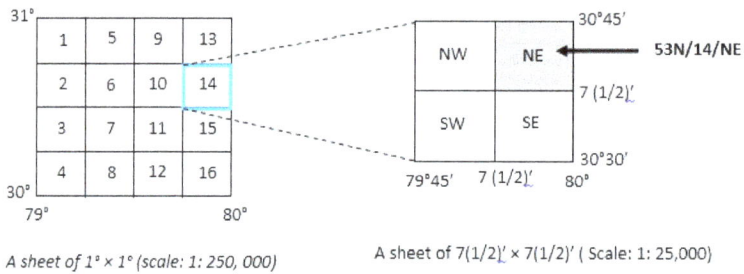

A sheet of 1° × 1° (scale: 1: 250, 000)   A sheet of 7(1/2)' × 7(1/2)' (Scale: 1: 25,000)

The classification of maps has already been discussed in unit 1, given below is a brief account of distribution maps.

Thematic Map

Edmond Halley's *New and Correct Chart Shewing the Variations of the Compass* (1701), the first chart to show lines of equal magnetic variation.

A thematic map is a type of map specifically designed to show a particular theme connected with a specific geographic area.

Overview

A *thematic map* is a map that focuses on a specific theme or subject area. This is in contrast to *general reference maps*, which regularly show the variety of phenomena—geological, geographical, political—together. The contrast between them lies in the fact that thematic maps use the base data, such as coastlines, boundaries and places, only as points of reference for the phenomenon being mapped. General maps portray the base data, such as landforms, lines of transportation, settlements, and political boundaries, for their own sake.

Thematic maps emphasize spatial variation of one or a small number of geographic distributions. These distributions may be physical phenomena such as climate or human characteristics such as population density and health issues. Barbara Petchenik described the difference as "in place, about space." While general reference maps where something is in space, thematic maps tell a story about that place (e.g., city map).

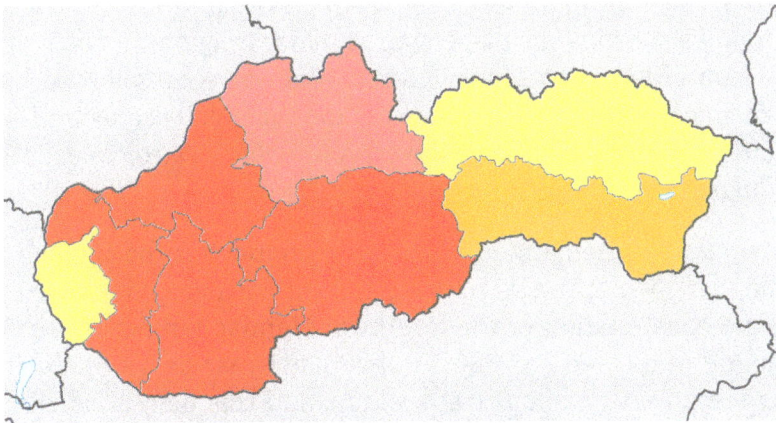

Map of the total fertility rate in Slovakia by region (2014)

1.5 - 1.7   1.4 - 1.5   1.3 - 1.4   < 1.3

Thematic maps are sometimes referred to as graphic essays that portray spatial variations and interrelationships of geographical distributions. Location, of course, is important to provide a reference base of where selected phenomena are occurring.

History

An important cartographic element preceding thematic mapping was the development of accurate base maps. Improvements in accuracy proceeded at a gradual pace, and even until the mid-17th century, general maps were usually of poor quality. Still, base maps around this time were good enough to display appropriate information, allowing for the first thematic maps to come into being.

John Snow's cholera map about the cholera deaths in London in the 1840s, published 1854.

One of the earliest thematic maps was a map entitled *Designatio orbis christiani* (1607) by Jodocus Hondius showing the dispersion of major religions, using map symbols in the French edition of his *Atlas Minor* (1607). This was soon followed by a thematic globe (in the form of a six-gore map) showing the same subject, using Hondius' symbols, by Franciscus Haraeus, entitled: *Novus typus orbis ipsus globus, ex Analemmate Ptolomaei diductus* (1614)

An early contributor to thematic mapping in England was the English astronomer Edmond Halley (1656–1742). His first significant cartographic contribution was a star chart of the constellation of the Southern Hemisphere, made during his stay on St. Helena and published on 1686. In that same year he also published his first terrestrial map in an article about trade winds, and this map is called the first meteorological chart. In 1701 he published the "New and Correct Chart Shewing the Variations of the Compass", the first chart to show lines of equal magnetic variation.

Another example of early thematic mapping comes from London physician John Snow. Though disease had been mapped thematically, Snow's cholera map in 1854 is the best known example of using thematic maps for analysis. Essentially, his technique and methodology anticipate principles of a geographic information system (GIS). Starting with an accurate base map of a London neighborhood which included streets and water pump locations, Snow mapped out the incidents of cholera death. The emerging pattern centered around one particular pump on Broad Street. At Snow's request, the handle of the pump was removed, and new cholera cases ceased almost at once. Further investigation of the area revealed the Broad Street pump was near a cesspit under the home of the outbreak's first cholera victim.

Another 19th century example of thematic maps, according to Friendly (2008), was the earliest known choropleth map in 1826 created by Charles Dupin. Based on this work Louis-Leger Gauthier (1815–1881) developed the population contour map, a map that shows the population density by contours or isolines.

Uses of Thematic Maps

Thematic maps serve three primary purposes.

1. They provide specific information about particular locations.

2. They provide general information about spatial patterns.

3. They can be used to compare patterns on two or more maps.

Common examples are maps of demographic data such as population density. When designing a thematic map, cartographers must balance a number of factors in order to effectively represent the data. Besides spatial accuracy, and aesthetics, quirks of human visual perception and the presentation format must be taken into account.

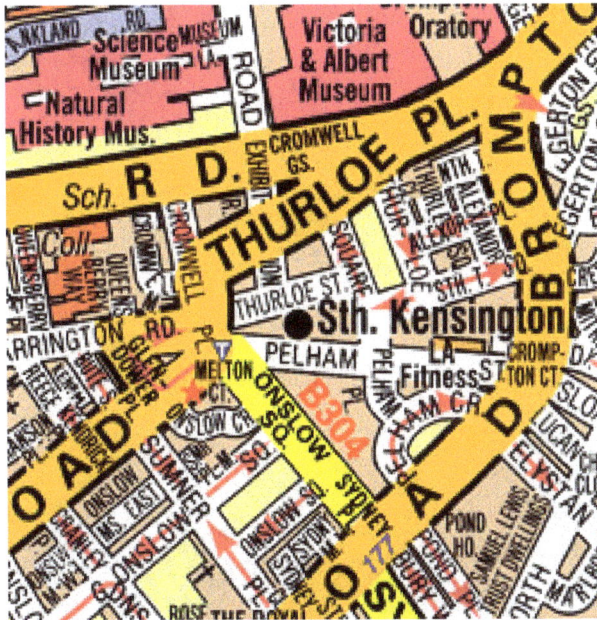

Section of a Street/City map from a Geographers' A-Z Map Company Atlas

In addition, the audience is of equal importance. Who will "read" the thematic map and for what purpose helps define how it should be designed. A political scientist might prefer having information mapped within clearly delineated county boundaries (choropleth maps). A state biologist could certainly benefit from county boundaries being on a map, but nature seldom falls into such smooth, man-made delineations. In which case, a dasymetric map charts the desired information underneath a transparent county boundary map for easy location referencing.

Data Terminology

A thematic map is univariate if the non-location data is all of the same kind. Population density, cancer rates, and annual rainfall are three examples of univariate data.

Bivariate mapping shows the geographical distribution of two distinct sets of data. For example, a map showing both rainfall and cancer rates may be used to explore a possible correlation between the two phenomena.

More than two sets of data leads to multivariate mapping. For example, a single map might show population density in addition to annual rainfall and cancer rates.

Methods of Thematic Mapping

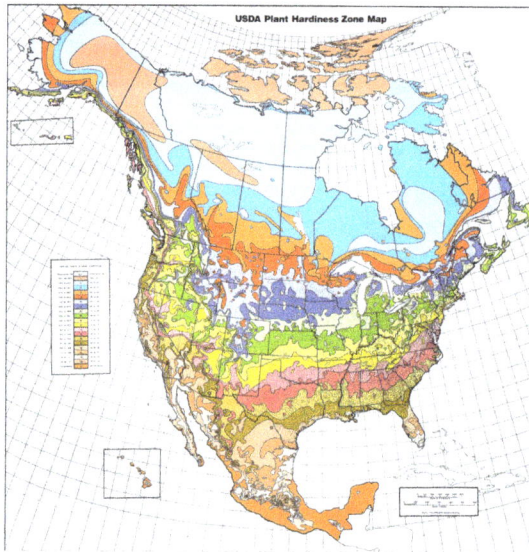

Map of climate and plant hardiness zones.

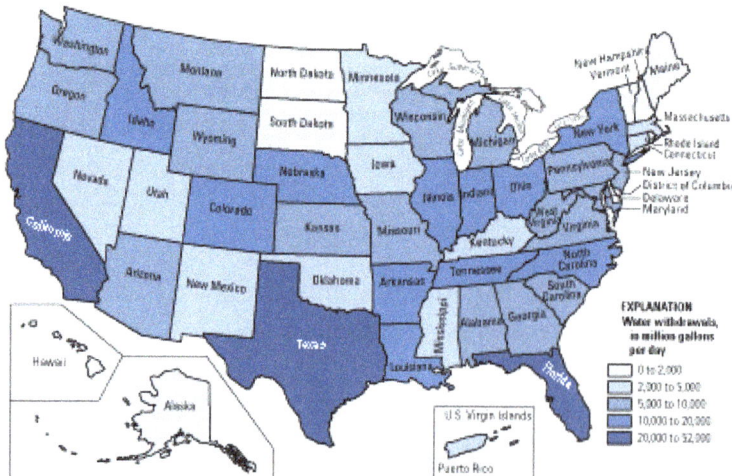

Choropleth map of water use.

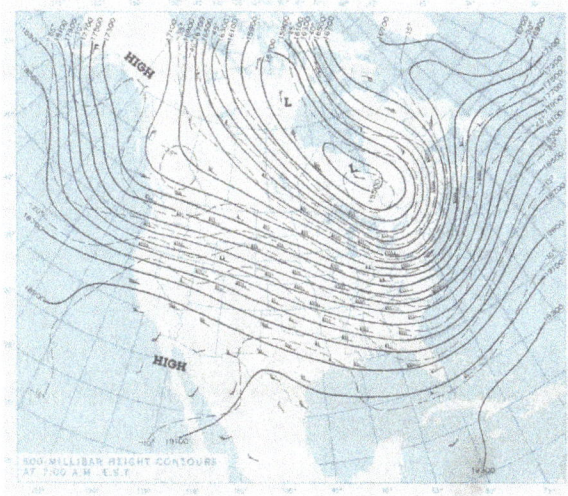

Isarithmic map of barometric pressure.

Cartographers use many methods to create thematic maps, but five techniques are especially noted.

Choropleth

Choropleth mapping shows statistical data aggregated over predefined regions, such as counties or states, by coloring or shading these regions. For example, countries with higher rates of infant mortality might appear darker on a choropleth map. This technique assumes a relatively even distribution of the measured phenomenon within each region. Generally speaking, differences in hue are used to indicate qualitative differences, such as land use, while differences in saturation or lightness are used to indicate quantitative differences, such as population.

Proportional Symbol

The proportional symbol technique uses symbols of different sizes to represent data associated with different areas or locations within the map. For example, a disc may be shown at the location of each city in a map, with the area of the disc being proportional to the population of the city.

Isarithmic or Isopleth

Isarithmic maps, also known as contour maps or isopleth maps depict smooth continuous phenomena such as precipitation or elevation. Each line-bounded area on this type of map represents a region with the same value. For example, on an elevation map, each elevation line indicates an area at the listed elevation. An Isarithmic map is a planimetric graphic representation of a 3-D surface. Isarithmic mapping requires 3-D thinking for surfaces that vary spatially.

Dot

A dot distribution map might be used to locate each occurrence of a phenomenon, as in the map made by Dr. Snow during the 1854 Broad Street cholera outbreak, where each dot represented one death due to cholera. Where appropriate, the dot distribution technique may also be used in combination the proportional symbol technique

Dasymetric

A dasymetric map is an alternative to a choropleth map. As with a choropleth map, data are collected by enumeration units. But instead of mapping the data so that the region appears uniform, *ancillary information* is used to model internal distribution of the phenomenon. For example, population density will be much lower in forested area than urbanized area, so in a common operation, land cover data (forest, water, grassland, urbanization) may be used to model the distribution of population reported by census enumeration unit such as a tract or county.

Distribution Maps

These are the maps that depict distribution of objects having definite values. These can further be divided on the basis of method of construction. While constructing the maps the data can be presented by:

Color: A map which shows different objects using various colors is known as chorochromatic map. For example, the districts of a state can be depicted using multiple colors or shades of a single color.

District Map of Himachal Pradesh

Symbol: A map which uses symbols for representing the data is called choroschematic map. For example distribution of crop types in an area where rice is shown by symbol R, wheat by W, maize by M etc.

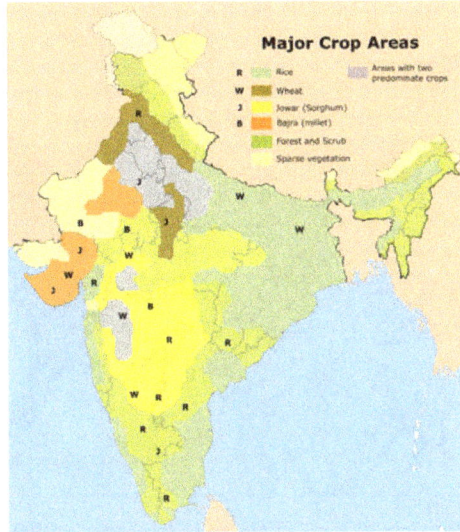

Regular lines: A map in which statistical data can be shown using lines of equal in-terval is called isopleth map. The lines are drawn to show equal amount of rainfall, pressure etc. The following figure shows economic potential isopleths for England and Wales.

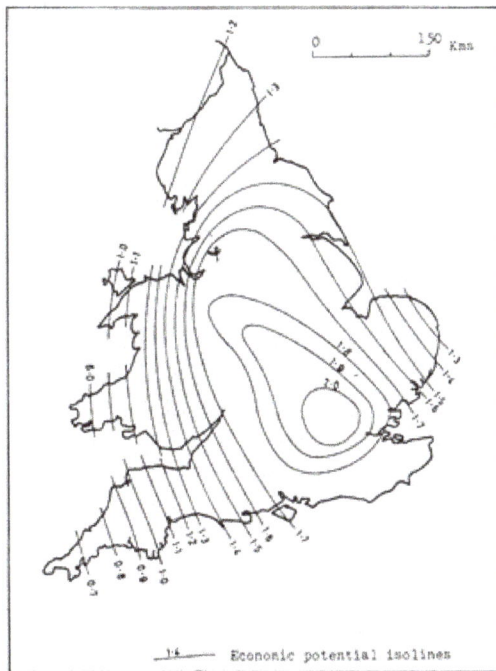

Dots: A map by which distribution of objects is shown by putting dots where each dot refers to a fixed number or quantity is called a dot map. For example the population distribution in a city can be shown using dot maps.

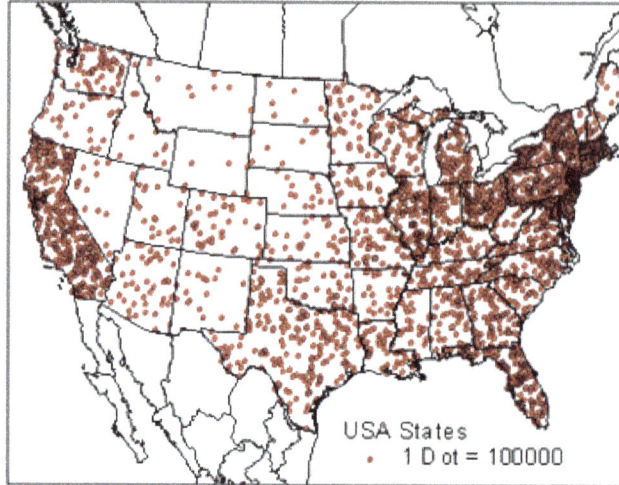

Shading: A map that uses different patterns of shading in order to represent the values of some property in an area. For example, the variation in production of fish in different states of a country can be shown using different patterns of shading.

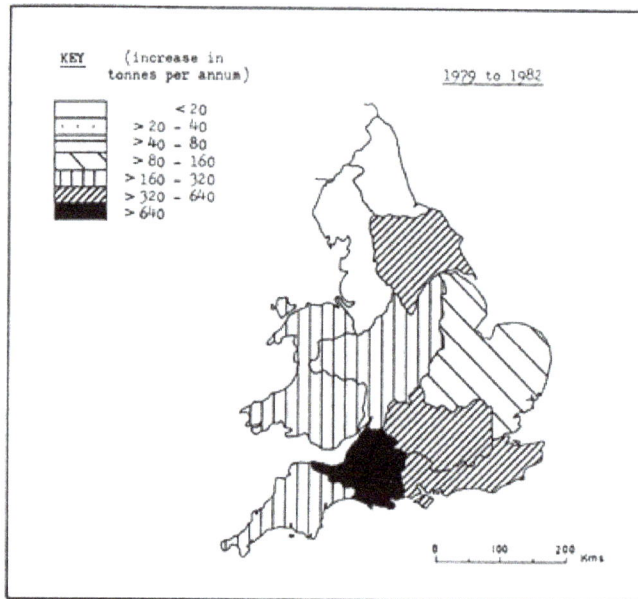

References

- Kurt A. Raaflaub; Richard J. A. Talbert (2009). Geography and Ethnography: Perceptions of the World in Pre-Modern Societies. John Wiley & Sons. p. 147. ISBN 1-4051-9146-5

- Catherine Delano Smith (1996). "Imago Mundi's Logo the Babylonian Map of the World". Imago Mundi. 48: 209–211. JSTOR 1151277. doi:10.1080/03085699608592846

- Needham, Joseph (1971). Part 3: Civil Engineering and Nautics. Science and Civilization in China. 4. Cambridge University Press. p. 569. ISBN 978-0-521-07060-7

- Sircar, D. C. C. (1990). Studies in the Geography of Ancient and Medieval India. Motilal Banarsidass Publishers. p. 330. ISBN 81-208-0690-5

- Bartz Petchenik, Barbara (April 1979). "From Place to Space: The Psychological Achievement of Thematic Mapping". Cartography and Geographic Information Science. 6 (1): 5–12. doi:10.1559/152304079784022763

GIS: Database Analysis and Management

Spatial database is the database that stores and retrieves data that defi nes a geometric space. It represents simple objects like lines or polygons and also complex structures like topological coverages and Triangulated irregular networks (TINs). This section discusses the methods of GIS database and spatial analysis in a critical manner providing key analysis to the subject matter.

Spatial DBMS

Computers and Information Technology have driven the human society towards a comfortable life where thousands of activities are performed on a click of a button and help save a lot of time. Everyday activities such as withdrawing money from a bank, making airline reservations, accessing a book from a library etc., are the processes that involve computer programs that access databases and database systems to facilitate all these services to the users.

The database systems have been growing and are involved in diverse applications such as:

- A multimedia database to store pictures, video clips, and sound messages.

- A real time database technology to control industrial and manufacturing processes.

- A GIS database used for storing the geographical information.

- Other common applications of databases include bank transactions, sales, reservations, registration etc.

Let us start understanding the world of databases with its functional unit known as data.

Data is defined as a collection of observations that could be in forms of facts, values, measurement, images etc. The data when used with a context becomes the information. It can be further processed, organized or simply summarized.

Given below is a list of numbers. This list of numbers is data since it communicates no meaning as such.

25
23
24
27

As soon as it is qualified that the numbers refer to the temperature recorded in degree Celsius in different cities of India on a particular day, the numbers become the temperature information.

Date	Cities	Temperature (°C)
23 October 2011	Delhi	25
	Mumbai	23
	Kolkata	24
	Chennai	27

A spatial database is a database that is optimized to store and query data that represents objects defined in a geometric space. Most spatial databases allow representing simple geometric objects such as points, lines and polygons. Some spatial databases handle more complex structures such as 3D objects, topological coverages, linear networks, and TINs. While typical databases have developed to manage various numeric and character types of data, such databases require additional functionality to process spatial data types efficiently, and developers have often added *geometry* or *feature* data types. The Open Geospatial Consortium developed the Simple Features specification (first released in 1997) and sets standards for adding spatial functionality to database systems. The *SQL/MM Spatial* ISO/EIC standard is a part the SQL/MM multimedia standard and extends the Simple Features standard with data types that support circular interpolations.

Geodatabase

A geodatabase (also geographical database and geospatial database) is a database of geographic data, such as countries, administrative divisions, cities, and related information. Such databases can be useful for websites that wish to identify the locations of their visitors for customization purposes.

Features of Spatial Databases

Database systems use indexes to quickly look up values and the way that most databases index data is not optimal for spatial queries. Instead, spatial databases use a spatial index to speed up database operations.

In addition to typical SQL queries such as SELECT statements, spatial databases can perform a wide variety of spatial operations. The following operations and many more are specified by the Open Geospatial Consortium standard:

- Spatial Measurements: Computes line length, polygon area, the distance between geometries, etc.

- Spatial Functions: Modify existing features to create new ones, for example by providing a buffer around them, intersecting features, etc.

- Spatial Predicates: Allows true/false queries about spatial relationships between geometries. Examples include "do two polygons overlap" or 'is there a residence located within a mile of the area we are planning to build the landfill?'

- Geometry Constructors: Creates new geometries, usually by specifying the vertices (points or nodes) which define the shape.

- Observer Functions: Queries which return specific information about a feature such as the location of the center of a circle

Some databases support only simplified or modified sets of these operations, especially in cases of NoSQL systems like MongoDB and CouchDB.

Spatial Index

Spatial indices are used by spatial databases (databases which store information related to objects in space) to optimize spatial queries. Conventional index types do not efficiently handle spatial queries such as how far two points differ, or whether points fall within a spatial area of interest. Common spatial index methods include:

- HHCode

- Grid (spatial index)

- Z-order (curve)

- Quadtree

- Octree

- UB-tree

- R-tree: Typically the preferred method for indexing spatial data. Objects (shapes, lines and points) are grouped using the minimum bounding rectangle (MBR). Objects are added to an MBR within the index that will lead to the smallest increase in its size.

- R+ tree

- R* tree

- Hilbert R-tree

- X-tree

- kd-tree

- m-tree – an m-tree index can be used for the efficient resolution of similarity queries on complex objects as compared using an arbitrary metric.

- Point access method

- Binary space partitioning (BSP-Tree): Subdividing space by hyperplanes.

Spatial Database Systems

- All OpenGIS specifications compliant products

- Open-source spatial databases and APIs, some of which are OpenGIS-compliant

- Caliper extends the Raima Data Manager with spatial datatypes, functions, and utilities.

- Boeing's Spatial Query Server spatially enables Sybase ASE.

- Smallworld VMDS, the native GE Smallworld GIS database

- SpatiaLite extends Sqlite with spatial datatypes, functions, and utilities.

- IBM DB2 Spatial Extender can spatially-enable any edition of DB2, including the free DB2 Express-C, with support for spatial types

- ClusterPoint offers native indexed support for distances, range matching and polygon matching, as well as aggregation.

- Oracle Spatial

- Oracle Locator

- Vertica Place, the geo-spatial extension for HP Vertica, adds OGC-compliant spatial features to the relational column-store database.

- Microsoft SQL Server has support for spatial types since version 2008

- PostgreSQL DBMS (database management system) uses the spatial extension PostGIS to implement the standardized datatype *geometry* and corresponding functions.

- Teradata Geospatial includes 2D spatial functionality (OGC-compliant) in its data warehouse system.

- MonetDB/GIS extension for MonetDB adds OGS Simple Features to the relational column-store database.

- Linter SQL Server supports spatial types and spatial functions according to the

OpenGIS specifications.

- MySQL DBMS implements the datatype *geometry*, plus some spatial functions implemented according to the OpenGIS specifications. However, in MySQL version 5.5 and earlier, functions that test spatial relationships are limited to working with minimum bounding rectangles rather than the actual geometries. MySQL versions earlier than 5.0.16 only supported spatial data in MyISAM tables. As of MySQL 5.0.16, InnoDB, NDB, BDB, and ARCHIVE also support spatial features.

- Neo4j – a graph database that can build 1D and 2D indexes as B-tree, Quadtree and Hilbert curve directly in the graph

- AllegroGraph – a graph database which provides a novel mechanism for efficient storage and retrieval of two-dimensional geospatial coordinates for Resource Description Framework data. It includes an extension syntax for SPARQL queries.

- MarkLogic, MongoDB, RavenDB, and RethinkDB support geospatial indexes in 2D.

- Esri has a number of both single-user and multiuser geodatabases.

- SpaceBase, a real-time spatial database.

- CouchDB a document-based database system that can be spatially enabled by a plugin called Geocouch

- CartoDB, a cloud-based geospatial database on top of PostgreSQL with PostGIS

- StormDB, an upcoming cloud-based database on top of PostgreSQL with geospatial capabilities

- AsterixDB, an open-source big data management system with native geospatial capabilities

- Kinetica, a GPU-accelerated analytics database optimized for geospatial analytics on large datasets.

- SpatialDB by MineRP, the world's first open-standards (OGC) spatial database with spatial type extensions for the Mining Industry

- H2 supports geometry types and spatial indices as of version 1.3.173 (2013-07-28). An extension called H2GIS available on Maven Central gives full OGC Simple Features support.

- GeoMesa is a cloud-based spatio-temporal database built on top of Apache Accumulo and Apache Hadoop. GeoMesa supports full OGC Simple Features support and a GeoServer plugin.

- Ingres 10S and 10.2 include native comprehensive spatial support. Ingres includes the Geospatial Data Abstraction Library cross-platform spatial data translator.

- Tarantool supports geospatial queries with RTREE index.
- SAP HANA supports geospatial with SPS08 .
- Redis with the Geo API

Data Storage

There are two approaches of storing data:

1. File based
2. Database

File Based Approach

File-based system is a collection of application programs that perform services for the users with each program defining and managing its data.

A flat file is an ordinary file where records of the file do not contain any information to communicate the file structure or relationship among the records to the application which is using the file.

Types of File Structure

- Unordered files
- Ordered files
- Index files

Unordered Files

	Name	Roll-ID	Class	Address
Block1	Ajay Kumar			
	Biren Das			
	Dipak Raj			
Block2	Rina Sharma			
	Prabhat Vij			
	Nira Thakur			
Block3	Tashina Rai			
	Priya Kiran			
New record ▶	Roshan Lal			

Also known as heap files, the unordered sequential files have the basic type of organization where records are placed in the file in the order in which they are inserted i.e. new records are inserted at the end of the file.

Ordered Files

The records in such a file can be ordered based on the values of one of their fields known as ordering fields. If the ordering field is the field whose value are distinct for each individual entity of the file, then the field is known as ordering key. Reading the records in order of ordering key values becomes efficient as no sorting is required.

	Name	Roll-ID	Class	Address
Block1	Ajay Kumar			
	Biren Das			
	Dipak Raj			
Block2	Rina Sharma			
	Prabhat Vij			
	Nira Thakur			
Block3	Tashina Rai			
	Priya Kiran			

Index Files

Indexes are additional access structures which are used to speed up the retrieval of the records in response to a search condition. These provide alternative ways of accessing the records without affecting the physical placement of records.

Data Structure Models

Data models are the conceptual models that describe the structures of databases. Structure of a database is defined by the data types, the constraints and the relationships for the description or storage of data. Following are the most often used data models:

Hierarchical Data Structure Model

It is the earliest database model that is evolved from file system where records are arranged in a hierarchy or as a tree structure. Records are connected through pointers that store the address of the related record. Each pointer establishes a parent child relationship where a parent can have more than one child but a child can only have one parent. There is no connection between the elements at the same level. To locate a particular record, you have to start at the top of the tree with a parent record and trace down the tree to the child.

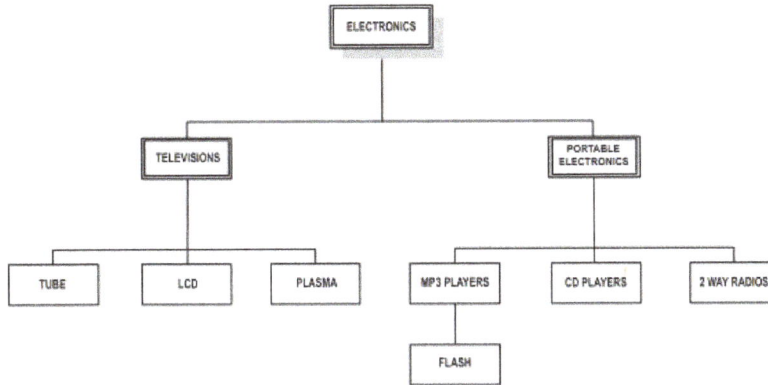

Example showing hierarchical data structure

The figure above describes the electronic gadgets in day today use. We can see that flash is a child of mp 3 players, which is a child of portable electronics, which is a child of electronics. The topmost element electronics has no parent.

Tube, LCD, plasma, CD players and 2 way radios are leaf nodes (don't have any children).

Advantages

- Easy to understand: The organization of database parallels a family tree understanding which is quite easy.

- Accessing records or updating records are very fast since the relationships have been predefined.

Disadvantages

- Large index files are to be maintained and certain attribute values are repeated many times which lead to data redundancy and increased storage.

- The rigid structure of this model doesn't allow alteration of tables, therefore to add a new relationship entire database is to be redefined.

Network Data Structure Model

A network is a generalized graph that captures relationships between objects using connectivity. A network database consists of a collection of records that are connected to each other through links. A link is an association between two records. It allows each record to have many parents and many children thus allowing a natural model of relationships between entities.

Advantages

- The many to many relationships are easily implemented in a network data model.

- Data access and flexibility in network model is better than that in hierarchical model. An application can access an owner record and the member records within a set .

- It enforces data integrity as a user must first define owner record and then the member records.

- The model eliminated redundancy but at the expense of more complicated relationships.

Disadvantages

The network model has a complex structure that requires familiarity from user's as well as programmer's end.

Relational Data Structure Model

The relational data model was introduced by Codd in 1970. The relational database relates or connects data in different files through the use of a common field. A flat file structure is used with a relational database model. In this arrangement, data is stored in different tables made up of rows and columns. The columns of a table are named by attributes. Each row in the table is called a tuple and represents a basic fact. No two rows of the same table may have identical values in all columns.

There are two crucial data integrity constraints viz. primary key and foreign key. A primary key is an attribute whose value is unique across all tuples (rows) in a relation (table). The primary key of one table appearing as an attribute of another table is known as a foreign key in that table.

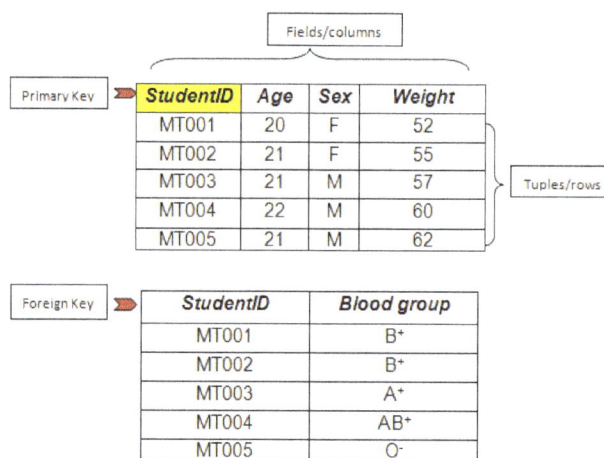

Fields/columns

Primary Key →

StudentID	Age	Sex	Weight
MT001	20	F	52
MT002	21	F	55
MT003	21	M	57
MT004	22	M	60
MT005	21	M	62

Tuples/rows

Foreign Key →

StudentID	Blood group
MT001	B+
MT002	B+
MT003	A+
MT004	AB+
MT005	O-

Advantages

- The manager or administrator does not have to be aware of any data structure or data pointer. One can easily add, update, delete or create records using simple logic.

Disadvantages

- A few search commands in a relational database require more time to process compared with other database models.

Working with Tables

Most databases instead of keeping their data together in a single table, organize the data into multiple tables each focusing on a specific topic. A user can link these tables if required information isn't present in a single table. The records in one table can be associated with records in the other table through a common field. The temporary associations can be made by joining and relating the tables.

Join

Joining appends the fields of one table to fields of another through an attribute/field common to both the tables.

StID	Age
N002	19
N005	21
N012	20
N015	19

Join Table

StID	Rank
N002	2
N005	3
N012	4
N015	1

Target

StID	Age	Rank
N002	19	2
N005	21	3
N012	20	4
N015	19	1

Relate

Relate defines a relationship between two tables. Relates are bidirectional which means both tables involved will be able to use the relate regardless of which table owns the relate. The associated data isn't appended in the table like it is in join. However one can access the related data by selecting a particular record and then going to the related tables against that record.

Database and Database Management System

A database is a collection of logically related data. It represents an aspect of a real world and is designed, built or populated with data for a specific purpose. Many databases exist for many applications, and each one of them is maintained by a collection of programs known as a database management system.

A database management system (DBMS) is a computer program that stores and manages large amounts of data. One can define, construct, edit and share the database among various users and applications.

A database is defined by the data types, structures, and data constraints that are stored in the database. Constructing the database means storing the data in the database.

Manipulating involves querying the database to retrieve specific data, updating the database, and generation of reports. Sharing is allowing multiple users and programs to access the database concurrently. Database contents are divided into two parts viz. schema and data. Schema is the structure of database. It indicates the rule which data must obey. Data on the other hand are the facts.

Imagine that we want to store the information about the students of a particular class. This information in a database would be stored together in a single container called table. The table has rows (with different students) and columns (that contain facts on the students such as studentname, age etc.) The table is named STUDENT_INFO.

STUDENT_INFO			
STUDENTNAME	STUDENTID	AGE	SEX
ALICE	CE01	18	F
ANDREW	CE02	17	M
DAVID	CE03	18	M
DONA	CE04	18	F

The schema would define that STUDENT_INFO has four facts/attributes viz. ' STUDENTNAME', 'STUDENTID', 'AGE', 'SEX'. To ensure that correct data is filled in all the columns of the table one can also enforce rules and constraints for data input.

Advantages of DBMS

- Controlling Redundancy

  Redundancy means storing the same data multiple times. DBMS checks redundancy and prevents duplication of efforts, saves storage space and preserves the data files from becoming inconsistent.

- Restricting Unauthorized Access

  A DBMS provides a security and authorization system, which the database administrator uses to create accounts and to specify account restriction.

- Providing Storage Structures for Efficient Query Processing

  Database systems provide capabilities for efficient execution of queries and updates. Because the database is typically stored on disk, it provides specialized data structures to speed up disk search for the desired records. Auxiliary files called indexesare used for this purpose.

- Providing Backup and Recovery

  The backup and recovery subsystem of the DBMS helps in recovering from hardware or software failures.

- Enforcing Integrity Constraints

    Most database applications have certain integrity constraints that must be held for the data. A DBMS should provide capabilities for defining and enforcing these constraints.

Along with the advantages, the DBMS usage involves overhead costs that are not incurred in conventional file processing. These overhead costs are due to:

- High initial investment in hardware, software and training

- Overhead for providing security, concurrency control, recovery and integrity functions

One may use regular files instead of a DBMS under the following circumstances:

- The database and applications are simple, well defined and are not expected to change.

- Multiple user access to data isn't required

Database Architecture

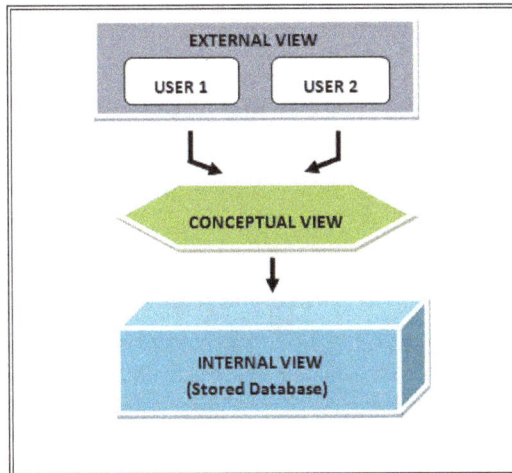

Three tier architecture of a database

We can consider the database on three levels of abstraction:

1. *External Level* refers to user's view of the database. It describes a part of the database for particular group of users. Depending on their needs, different users access different parts of the database. It employs a powerful and flexible security mechanism by hiding parts of the database from certain users.

2. *Conceptual Level* refers to the logical structure of the entire database. It describes data as well as the relationships among the data.

3. *Internal Level* refers to the details of physical storage of the database on the computer. It consists of description of storage space allocation for data and indexes, record placements and data compression.

Normalization

Normalization is a design technique which helps designing relational databases. The objective of normalization is to create a set of relational tables that are free of redundant data and to make data consistent.

First normal form

- Eliminate repeating groups in individual tables.

- Create a separate table for each set of related data.

- Identify each set of related data with a primary key.

Second normal form

- Create separate tables for sets of values that apply to multiple records .

- Relate these tables with a foreign key

Third normal form

- Eliminate fields that do not depend on the key

An example explains the process of normalization which is described as following:

Unnormalized Table

Table1: Student

Student#	Advisor	Adv-Room	Class1	Class2	Class3
1022	Jones	412	101-07	143-01	159-02
4123	Smith	216	201-01	211-02	214-01

The example shows a table (Table1) in which the group 'class' is mentioned three times. Since one student has several classes, these classes should be listed in a separate table. Another table in first normal form (Table2) is created by eliminating the repeating group (Class#), as shown below:

First Normalization (No Repeating Groups)

Note the Class# values for each Student# value in the above table. Class# is not functionally dependent on Student# (primary key), so this relationship is not in second normal form. Class# is separated from the first normalization table and is placed in another table (Table4).

Table2: Student

Student#	Advisor	Adv-Room	Class#
1022	Jones	412	101-07
1022	Jones	412	143-01
1022	Jones	412	159-02
4123	Smith	216	201-01
4123	Smith	216	211-02
4123	Smith	216	214-01

Second Normalization (Eliminate Redundant Data)

Table 3: Student

Student#	Advisor	Adv-Room
1022	Jones	412
4123	Smith	216

Table 4: Registration

Student#	Class#
1022	101-07
1022	143-01
1022	159-02
4123	201-01
4123	211-02
4123	214-01

The attribute Adv-Room is functionally dependent on the Advisor attribute. It must be moved from the student table to some other table let's say faculty table (Table6).

Third Normalization (Eliminate Data not Dependent on Key)

Table 5: Student

Student#	Advisor
1022	Jones
4123	Smith

Table 6: Faculty

Name	Room	Dept
Jones	412	42
Smith	216	42

Object Oriented Database

According to Worboys (1995) spatial data can't be managed properly by relational data model approach. Spatial data don't naturally fit into tabular structures. More attention has been given to the development of object oriented approach of database design. The aim of object oriented model is to allow data modeling which is closer to real world. An object-oriented database uses objects as elements within database files. An object is a logical grouping of related data that represents a real world entity. Each object is a distinct entity which is identified using a key attribute called ObjectID. The object can be grouped together to form a class. Objects of the same class have same attributes, behavior and relationships with other objects.

The attributes of the object are known as its states. In addition to the states of an object information about its behavior is stored. These are the operations that are performed on the object.

For example 'hotel' is an object with states such as name, address, number of rooms. The operations that can be performed on this object can vary as plotting on map, adding to a database, increasing or decreasing number of rooms etc. The object hotel belongs to a class called 'buildings'. Any object in the class hotel will inherit the properties of the class building. Thus the states and operations applicable to 'building' will also apply to 'hotel'.

Longley et al. (2001) list the following features of object data models which make them good for modeling GIS systems:

- Encapsulation: packaging together of the description of state and behavior in each object

- Inheritance: ability to use some or all characteristics of one object in another object

- Polymorphism: specific implementation of operations such as create, delete etc for each object.

Spatial Data Input & Editing

Collecting data and creating a GIS database is a time consuming but an important task. There are many sources of geographic data and many ways to enter that data into a GIS. A data pool can be generated by either data capture or data transfer. The data sources are divided into following two main classes:

Primary data

It involves direct measurement of objects and phenomena. Given below is the partial list of primary data:

Remote sensing data capture: Remote sensing refers to the technique of deriving the information about the objects without getting in physical contact with them. The information is derived from the measurements of the amount of electromagnetic (EM) radiations reflected, emitted or scattered from the objects under observation. The response is measured /captured by the sensors deployed in air or in space. The remote sensing data is often talked in terms of spatial, spectral and temporal resolutions.

Spatial resolution: It refers to the size of the object that can be resolved and is the measure of the pixel size.

Spectral resolution: It refers to the wavelengths of the EM spectrum in which response of the objects is captured.

Temporal resolution: It refers to the frequency with which data is captured for the same area.

Aerial photographic data is as important as remote sensing data for a GIS project. Though both aerial photographs and remote sensing images are technically similar, they have few differences as well. The most notable difference is that aerial photographs are captured using analog optical cameras and are then rasterized by scanning a film negative. Now a days digital cameras are being used for aerial photography. The aerial photographs are suitable for surveying and mapping projects.

Both satellite images and aerial photographs can provide stereo imagery from overlapping pairs of images i.e. they can generate a three dimensional model of the earth's surface. The other advantages include global coverage and repetitive monitoring that make these datasets useful for large area projects and short time events.

Surveying: Ground surveying is based on the principle of determining the 3D location of a point with the help of angles and distance measured from other known points. Survey starts from a benchmark position. The location of all surveyed points is relative to other points. The traditional surveying involves the use of transits, theodolites, chains and tapes for angle and distance measurement. These days, electro-optical devices called total stations measure both angles as well distance to an accuracy of 1mm. Surveying is a time and resource consuming activity but is the best way of obtaining accurate geographical data.

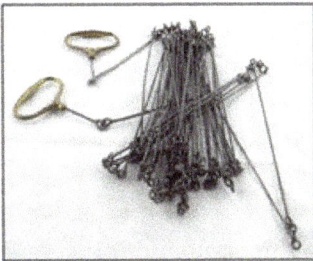  

(a) Chain   (b) Theodolite   (c) Long Tape

(d) Transit Level   (e) Total station

Survey Instruments

Sampling

Since it is not practically possible as well as worthwhile to observe the value of a variable at every point throughout the study area we adopt the strategy of sampling. Using sampling we measure subsets of the features in the area that best capture the spatial variation of the concerned attribute over the study area. The following five patterns options may be considered for sampling:

a. Simple random

This method ensures that all parts of the project area have an equal chance of being sampled. Project area is divided into a grid with numbered coordinates. A random site is picked by selecting coordinate pairs from a number table and plotting those on the project area map. Each random site is a sample point.

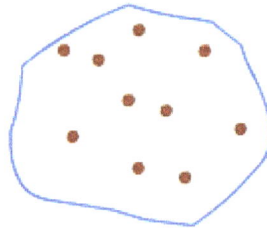

Simple random pattern

Advantage

- Operator bias is minimal

Drawback

- Classes with small areas may be inadequately sampled or missed entirely.

- Some of the sample points may be inaccessible on ground.

b. Stratified random

It maintains randomness and at the same time overcomes the chance of an uneven distribution of points among the map classes. Specific numbers of sample points are assigned to each class with respect to its size and significance for the project. Within a class the random sites are generated in the same way as in simple random pattern.

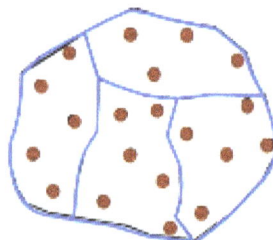

Stratified random pattern

c. Systematic

It arranges sample points at equidistant intervals thus forming a grid. Orientation of the grid is chosen randomly.

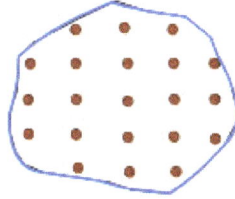

Systematic pattern

Drawback

• Randomness is not achieved because position of every sample point is determined by the choice of the starting point.

d. Systematic unaligned

It distributes the project area into a grid and assigns the positions of sample points randomly within the grid cells.

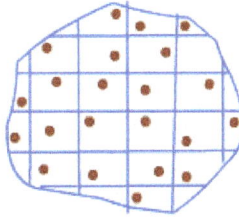

Systematic unaligned pattern

Advantages

• All parts of the project area are sampled.

• Randomness is maintained within the grid cells.

e. Clustered

In this method, nodal points are the centers for clusters of sample points. The nodal locations are selected randomly, stratified by classes, or by identification of accessible sites.

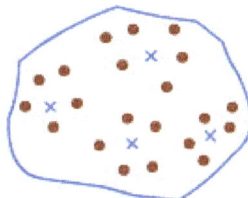

Clustered pattern

Advantage

- In terrain with poor access, the operator can make the most of accessible sites.

- Field time is greatly reduced as lesser sites are to be visited.

GPS (Global Positioning System): GPS is a collection of 27 NAVSTAR satellites orbiting the earth at a height of 12, 500 miles or 20, 200 km. It was originally funded by US Department of Defense and was only used for military purposes but, in the year 2000 it was opened for civilians as well. GPS works on distance and time principal. The GPS satellites transmit radio signals that indicate their exact position in space. The receiver measures the time taken by signal to reach the receiver. Similarly, distance from three or more satellites helps in triangulating the position of the receiver on the earth's surface. As soon as the signal from fourth satellite is received, elevation information is also derived. GPS has led to the development of hundreds of applications affecting every aspect of modern day to day life. Farming, mining, construction, logistics, communication, power etc. are some of the sectors that have started depending heavily on GPS.

Secondary Data

It refers to the data obtained from maps, hardcopy documents etc. Some of the methods to capture secondary data are as follows:

Scanned data: A scanner is used to convert analog source map or document into digital images by scanning successive lines across a map or document and recording the amount of light reflected from the data source. Documents such as building plans, CAD drawings, images and maps are scanned prior to vectorization. Scanning helps in reducing wear and tear; improves access and provides integrated storage.

There are three different types of scanner that are widely used:

1. Flat bed scanner

2. Rotating drum scanner

3. Large format feed scanner

(a) Flat bed scanner   (b) Rotating drum scanner   (c) Large format feed scanner

Flat bed scanner is a PC peripheral which is small and comparatively inaccurate. The rotating drum scanners are accurate but they tend to be slow and expensive. Large format feed scanner are the most suitable type for inputting GIS data as they are cheap, quick and accurate.

Digitization: Digitizing is the process of interpreting and converting paper map or image data to vector digital data.

Heads Down Digitization

Digitizers are used to capture data from hardcopy maps. Heads down digitization is done on a digitizing table using a magnetic pen known as Puck. The position of a cursor or puck is detected when passed over a table inlaid with a fine mesh of wires. The function of a digitizer is to input correctly the coordinates of the points and the lines. Digitization can be done in two modes:

1. Point mode: In this mode, digitization is started by placing a point that marks the beginning of the feature to be digitized and after that more points are added to trace the particular feature (line or a polygon). The number of points to be added to trace the feature and the space interval between two consecutive points are decided by the operator.

2. Stream mode: In stream digitizing, the cursor is placed at the beginning of the feature, a command is then sent to the computer to place the points at either equal or unequal intervals as per the position of the cursor moving over the image of the feature.

Heads-up Digitization

This method uses scanned copy of the map or image and digitization is done on the screen of the computer monitor. The scanned map lays vertical which can be viewed without bending the head down and therefore is called as heads up digitization.

Heads down digitization

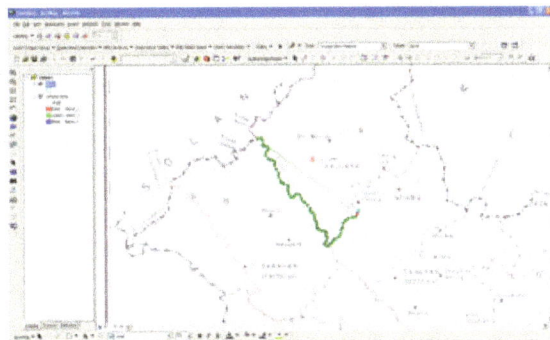

Screenshot of On-screen/Heads up digitization

Semi-automatic and automatic methods of digitizing requires post processing but saves lot of time and resources compared to manual method and is described in the following section.

Vectorization

Vectorization is the process of converting a raster image into a vector image. It is a faster way of creating the vector data from raster data. Automatic vectorization is performed in either batch or interactive mode. Batch vectorization takes one raster file and converts it into vector objects in a single operation. Post vectorization editing is required to remove the errors. In interactive vectorization software is used to automate digitizing. The operator snaps the cursor to a pixel and indicates the direction in which line is to be digitized. The software then automatically digitizes the line. The operator can decide various parameters such as density of points, whether to pause at junction for operator's intervention or to trace in a specific direction etc. Though the process involves labor it produces high quality data and greater productivity than the manual digitization.

Photogrammetry: It is the science of making measurements from aerial photographs and images. Apart from the 2D measurement from a single photograph, photogrammetry is also used for making 3D measurements from models made using stereo pairs of photographs. To make a 3D model, there must be 60% overlap along each flight line and 30% overlap between flight lines. The measurements from overlapping pairs of photographs are captured using stereoplotters. These build a model and allow 3D measurements to be captured, edited, stored and plotted. One can extract vector objects from 3D model in a way similar to the above discussed digitization.

Obtaining Data from external sources: Creating the same dataset multiple times for the same area is a time and resource intensive process. One can always import data from data repositories. Some of these are freely available while others are available at a price. Internet is the best way to search geographic data. The internet gives information about geographic data catalogs and vendors. National agencies of a state/country also disseminate geographic data through their web portals or through other digital media on demand made by the users.

Data Editing

Errors affect the quality of GIS data. Once the data is collected, and prepared for visualization and analysis it must be checked for errors.

Burrough (1986) divided the sources of error into the following categories:

1. Common sources of error

2. Errors resulting from original measurements

3. Errors arising through processing

Common Sources of Error

- Old data sources: The data sources used for a GIS project may be too old to use. Data collected in past may not be acceptable for current time projects.

- Lack of data: The data for a given area may be incomplete or entirely lacking. For example the land-use map for border regions may not be available.

- Map scale: The details shown on a map depend on the scale used. Maps or data of the appropriate scale at which details are required, must be used for the project. Use of wrong scale would make the analysis erroneous.

- Observation: High density of observations in an area increases the reliability of the data. Insufficient observations may not provide the level of resolution required for adequate spatial analysis as expected from the project.

Errors Resulting from Original Measurements

- Positional accuracy: Representing correct positions of geographic features on map depend upon the data being used. Biased field work, improper digitization and scanning errors result in accuracies in GIS projects.

- Content accuracy: Maps must be labeled correctly. An incorrect labeling can introduce errors which may go unnoticed by the user. Any omission from map or spatial database may result in inaccurate analysis.

Errors Arising through Processing

- Numerical errors: Different computers have different capabilities for mathematical operations. Computer processing errors occur in rounding off operations and are subject to the inherent limits of number manipulation by the processor.

- Topological errors: Data is subject to variation. Errors such as dangles, slivers, overlap etc are found to be present in the GIS data layers.

Dangle: An arc is said to be a dangling arc if either it is not connected to another arc properly (undershoot) or is digitized past its intersection with another arc (overshoot).

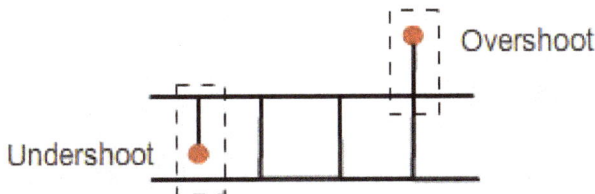

Sliver: It refers to the gap which is created between the two polygons when snapping is not considered while creating those polygons.

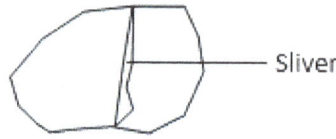

Sliver

These errors can be corrected using the constraints or the rules which are defined for the layers. Topology rules define the permissible spatial relationships between features.

- Digitizing and geocoding: Many errors arise at the time of digitization, geocoding, overlaying or rasterizing. The errors associated with damaged source maps and error while digitizing can be corrected by comparing original maps with digitized versions.

Raster data editing is concerned with correcting the specific contents of raster images than their general geometric characteristics. The objective of the editing is to produce an image suitable for raster geoprocessing. Following editing functions are mostly used for raster data editing:

- Filling holes and gaps: To fill holes and gaps that appear in the raster image

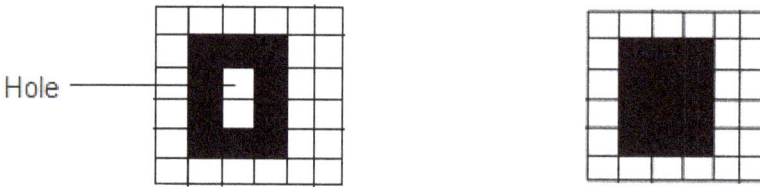

Hole

- Edge smoothing: To remove or fill single pixel irregularities in the foreground pixels and background pixels along lines

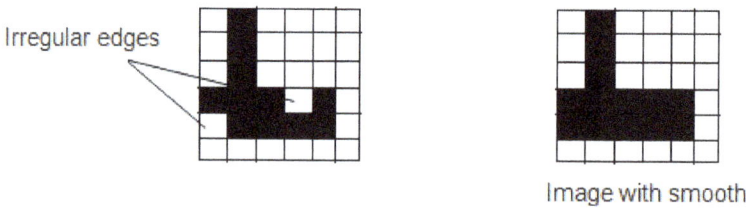

Irregular edges

Image with smooth

- Deskewing: To rotate the image by a small angle so that it is aligned orthogonally to the x and y axes of the computer screen

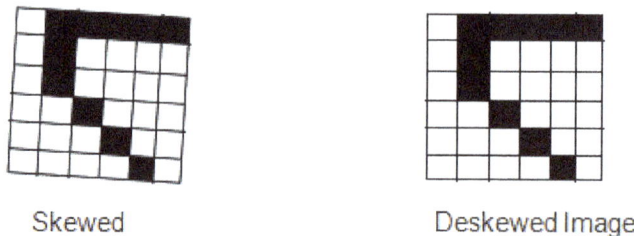

Skewed   Deskewed Image

- Filtering: To remove speckles or the random high or low valued pixels in the image

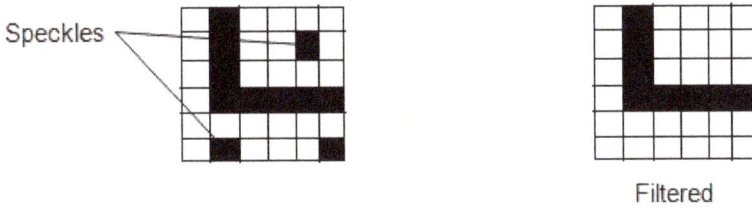

Speckles

Filtered

- Clipping and delete: To create a subset of an image or to remove unwanted pixels

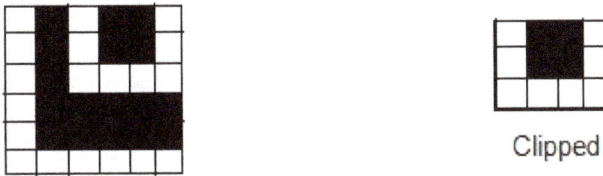

Clipped

Vector data editing is a post digitizing process that ensures that the data is free from errors. It suggests that

- Lines intersect properly without having any undershoots or overshoots

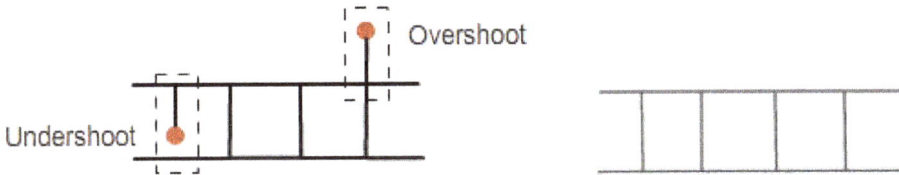

Overshoot

Undershoot

- Nodes are created at all points where lines intersect

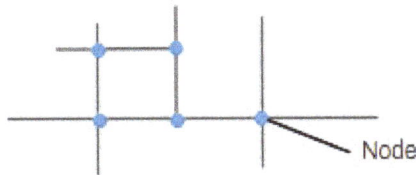

Node

- All polygons are closed and each of them contain a label point

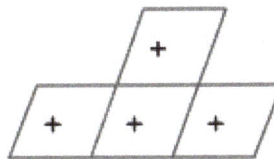

- Topology of the layer is built

Spatial Analysis

Spatial analysis is core to the GIS since all the hard work done in compiling and processing of the spatial data is to achieve this important outcome. It is only through spatial analysis that answers to a range of questions are found. Thus the spatial analysis helps in decision making.

The range of methods deployed for spatial analysis varies with respect to the type of the data model used. Before discussing these specific procedures pertaining to specific data models there are some issues that are generic and are discussed below.

Measurement: Measurement of length, perimeter and area of the features is a very common requirement in spatial analysis. However different methods are used to make measurements based on the type of data used i.e. vector or raster. Invariably, the measurements will not be exact, as digitized feature on map may not be entirely similar to the features on the ground, and moreover in the case of raster, the features are approximated using a grid cell representation.

In raster data model, the distance between two points can be calculated in the following ways:

a. Euclidean distance method: A straight line is drawn joining the two points and a right angled triangle is created. The distance is then derived using the Pythagorean geometry.

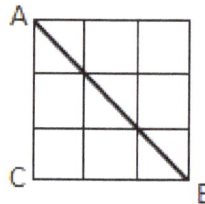

$$AB = \sqrt{AC^2 + CB^2}$$

If the cell size of the raster is 10m, then distance $AB = \sqrt{30^2 + 30^2}$

$$= \sqrt{1800}$$

$$= 42.42m$$

b. Manhattan distance method: In this method, distance along the raster cell sides from one point to the other is taken. The following example illustrates the method.

If the cell size of the raster is 10m, then distance $AB = \left(Aa+ab+bc+cd+de+eB\right)$

$$= \left(10+10+10+10+10+10\right)$$

$$= 60m$$

Perimeter is calculated by counting the number of cells in each side that is making the boundary of the feature and then multiplying the count by resolution (cell size) of the raster grid. All the sides of the feature are then added. Area is calculated by counting the number of cells making a feature and multiplying the count by the area of an individual grid.

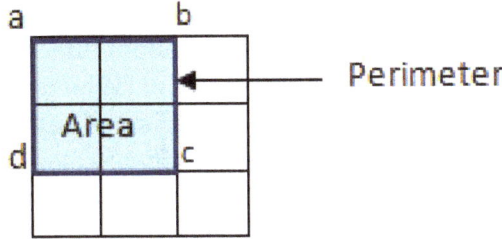

If the cell size of the raster is 10 m then the perimeter, P of the colored portion (abcd)

P=ab+bc+cd+ad

$$=(10+10)+(10+10)+(10+10)+(10+10)$$
$$=80m$$

Since abcd is made up of 4 cells each having an area of $(10\times10m^2)$, therefore its area is

$$A=4\times10\times10$$
$$=400m^2$$

In vector data model, distance between two points is measured using the Euclidean distance method. Perimeter is calculated by adding the measurements of straight lines forming the feature. To calculate the area, the feature is subdivided into geometric shapes and then the areas of the geometric shapes are totaled.

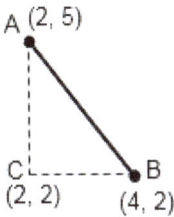

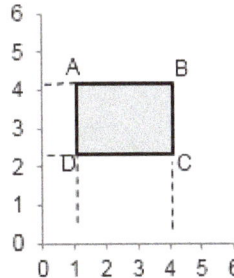

Area of ABCD=AB×AD
$$=3\times2$$
$$=6m^2$$

$AB = \sqrt{AC^2 + CB^2}$ Perimeter of ABCD=AB+BC+CD+AD
$$=\sqrt{9+4}$$ $$=3+2+3+2$$
$$=3.60m$$ $$=10m$$

Query

Query is a logical question which is performed on the database to retrieve specific data. Queries are useful for checking the quality of the data and the results obtained. There are two types of queries that can be performed in GIS:

- Aspatial or attribute queries: questions about the attributes of the feature. These do not include any spatial information. "Who owns the Star coffee shop?" is a simple query that does not involve analysis of any spatial component. Such queries could be performed by database software alone.

- Spatial queries: It involves selection of features based on location or other spatial information.

Where do the coffee shops with the same name lie in the city? Since the question asks for the location of coffee shops, the GIS software is able to show their locations on the digital map of the city.

Two or more queries can be combined together to identify features of interest. Boolean operators such as AND, OR, NOT, and XOR are used to combine queries. The spatial operations can differ depending on the data model used. The spatial operations pertaining to the vector and raster models are described below.

Map by Dr. John Snow of London, showing clusters of cholera cases in the 1854 Broad Street cholera outbreak. This was one of the first uses of map-based spatial analysis.

Spatial analysis or spatial statistics includes any of the formal techniques which study entities using their topological, geometric, or geographic properties. Spatial analysis includes a variety of techniques, many still in their early development, using different analytic approaches and applied in fields as diverse as astronomy, with its studies of the placement of galaxies in the cosmos, to chip fabrication engineering, with its use of "place and route" algorithms to build complex wiring structures. In a more restricted sense, spatial analysis is the technique applied to structures at the human scale, most notably in the analysis of geographic data.

Complex issues arise in spatial analysis, many of which are neither clearly defined nor completely resolved, but form the basis for current research. The most fundamental of these is the problem of defining the spatial location of the entities being studied.

Classification of the techniques of spatial analysis is difficult because of the large number of different fields of research involved, the different fundamental approaches which can be chosen, and the many forms the data can take.

History of Spatial Analysis

Spatial analysis can perhaps be considered to have arisen with early attempts at cartography and surveying but many fields have contributed to its rise in modern form. Biology contributed through botanical studies of global plant distributions and local plant locations, ethological studies of animal movement, landscape ecological studies of vegetation blocks, ecological studies of spatial population dynamics, and the study of biogeography. Epidemiology contributed with early work on disease mapping, notably John Snow's work of mapping an outbreak of cholera, with research on mapping the spread of disease and with location studies for health care delivery. Statistics has contributed greatly through work in spatial statistics. Economics has contributed notably through spatial econometrics. Geographic information system is currently a major contributor due to the importance of geographic software in the modern analytic toolbox. Remote sensing has contributed extensively in morphometric and clustering analysis. Computer science has contributed extensively through the study of algorithms, notably in computational geometry. Mathematics continues to provide the fundamental tools for analysis and to reveal the complexity of the spatial realm, for example, with recent work on fractals and scale invariance. Scientific modelling provides a useful framework for new approaches.

Fundamental Issues in Spatial Analysis

Spatial analysis confronts many fundamental issues in the definition of its objects of study, in the construction of the analytic operations to be used, in the use of computers for analysis, in the limitations and particularities of the analyses which are known, and in the presentation of analytic results. Many of these issues are active subjects of modern research.

Common errors often arise in spatial analysis, some due to the mathematics of space, some due to the particular ways data are presented spatially, some due to the tools which are available. Census data, because it protects individual privacy by aggregating data into local units, raises a number of statistical issues. The fractal nature of coastline makes precise measurements of its length difficult if not impossible. A computer software fitting straight lines to the curve of a coastline, can easily calculate the lengths of the lines which it defines. However these straight lines may have no inherent meaning in the real world, as was shown for the coastline of Britain.

These problems represent a challenge in spatial analysis because of the power of maps as media of presentation. When results are presented as maps, the presentation combines spatial data which are generally accurate with analytic results which may be inaccurate, leading to an impression that analytic results are more accurate than the data would indicate.

Spatial Characterization

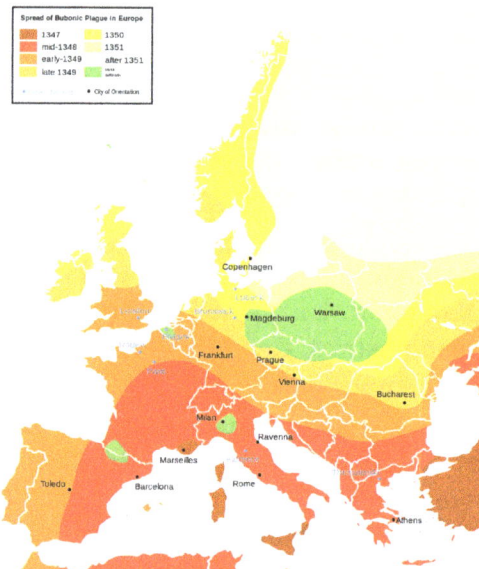

Spread of bubonic plague in medieval Europe. The colors indicate the spatial distribution of plague outbreaks over time.

The definition of the spatial presence of an entity constrains the possible analysis which can be applied to that entity and influences the final conclusions that can be reached. While this property is fundamentally true of all analysis, it is particularly important in spatial analysis because the tools to define and study entities favor specific characterizations of the entities being studied. Statistical techniques favor the spatial definition of objects as points because there are very few statistical techniques which operate directly on line, area, or volume elements. Computer tools favor the spatial definition of objects as homogeneous and separate elements because of the limited number of database elements and computational structures available, and the ease with which these primitive structures can be created.

Spatial Dependency or Auto-correlation

Spatial dependency is the co-variation of properties within geographic space: characteristics at proximal locations appear to be correlated, either positively or negatively. Spatial dependency leads to the spatial autocorrelation problem in statistics since, like temporal autocorrelation, this violates standard statistical techniques that assume independence among observations. For example, regression analyses that do not compensate for spatial dependency can have unstable parameter estimates and yield unreliable significance tests. Spatial regression models capture these relationships and do not suffer from these weaknesses. It is also appropriate to view spatial dependency as a source of information rather than something to be corrected.

Locational effects also manifest as spatial heterogeneity, or the apparent variation in a process with respect to location in geographic space. Unless a space is uniform and boundless, every location will have some degree of uniqueness relative to the other locations. This affects the spatial dependency relations and therefore the spatial process. Spatial heterogeneity means that overall parameters estimated for the entire system may not adequately describe the process at any given location.

Scaling

Spatial measurement scale is a persistent issue in spatial analysis; more detail is available at the modifiable areal unit problem (MAUP) topic entry. Landscape ecologists developed a series of scale invariant metrics for aspects of ecology that are fractal in nature. In more general terms, no scale independent method of analysis is widely agreed upon for spatial statistics.

Sampling

Spatial sampling involves determining a limited number of locations in geographic space for faithfully measuring phenomena that are subject to dependency and heterogeneity. Dependency suggests that since one location can predict the value of another location, we do not need observations in both places. But heterogeneity suggests that this relation can change across space, and therefore we cannot trust an observed degree of dependency beyond a region that may be small. Basic spatial sampling schemes include random, clustered and systematic. These basic schemes can be applied at multiple levels in a designated spatial hierarchy (e.g., urban area, city, neighborhood). It is also possible to exploit ancillary data, for example, using property values as a guide in a spatial sampling scheme to measure educational attainment and income. Spatial models such as autocorrelation statistics, regression and interpolation can also dictate sample design.

Common Errors in Spatial Analysis

The fundamental issues in spatial analysis lead to numerous problems in analysis including bias, distortion and outright errors in the conclusions reached. These issues are often interlinked but various attempts have been made to separate out particular issues from each other.

Length

In a paper by Benoit Mandelbrot on the coastline of Britain it was shown that it is inherently nonsensical to discuss certain spatial concepts despite an inherent presumption of the validity of the concept. Lengths in ecology depend directly on the scale at which they are measured and experienced. So while surveyors commonly measure the length of a river, this length only has meaning in the context of the relevance of the measuring technique to the question under study.

o   Britain measured using a long yardstick

o   Britain measured using a medium yardstick

o   Britain measured using a short yardstick

Locational Fallacy

The locational fallacy refers to error due to the particular spatial characterization chosen for the elements of study, in particular choice of placement for the spatial presence of the element.

Spatial characterizations may be simplistic or even wrong. Studies of humans often reduce the spatial existence of humans to a single point, for instance their home address. This can easily lead to poor analysis, for example, when considering disease transmission which can happen at work or at school and therefore far from the home.

The spatial characterization may implicitly limit the subject of study. For example, the spatial analysis of crime data has recently become popular but these studies can only describe the particular kinds of crime which can be described spatially. This leads to many maps of assault but not to any maps of embezzlement with political consequences in the conceptualization of crime and the design of policies to address the issue.

Atomic Fallacy

This describes errors due to treating elements as separate 'atoms' outside of their spatial context. The fallacy is about transferring individual conclusions to spatial units.

Ecological Fallacy

The ecological fallacy describes errors due to performing analyses on aggregate data when trying to reach conclusions on the individual units. Errors occur in part from spatial aggregation. For example, a pixel represents the average surface temperatures within an area. Ecological fallacy would be to assume that all points within the area have the same temperature. This topic is closely related to the modifiable areal unit problem.

Solutions to the Fundamental Issues

Geographic Space

A mathematical space exists whenever we have a set of observations and quantitative measures of their attributes. For example, we can represent individuals' incomes or years of education within a coordinate system where the location of each individual can be specified with respect to both dimensions. The distances between individuals within this space is a quantitative measure of their differences with respect to income and education. However, in spatial analysis we are concerned with specific types of mathematical spaces, namely, geographic space. In geographic space, the observations correspond to locations in a spatial measurement framework that captures their proximity in the real world. The locations in a spatial measurement framework often represent locations on the surface of the Earth, but this is not strictly necessary. A spatial measurement framework can also capture proximity with respect to, say, interstellar space or within a biological entity such as a liver. The fundamental tenet is Tobler's First Law of Geography: if the interrelation between entities increases with proximity in the real world, then representation in geographic space and assessment using spatial analysis techniques are appropriate.

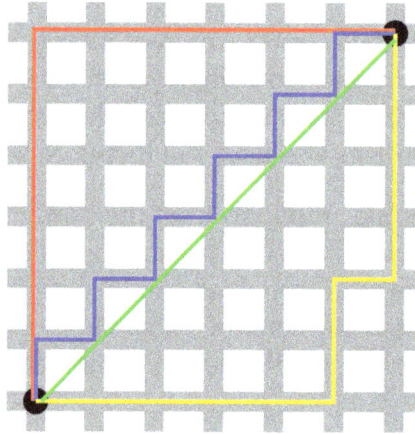

Manhattan distance versus Euclidean distance: The red, blue, and yellow lines have the same length (12) in both Euclidean and taxicab geometry. In Euclidean geometry, the green line has length $6 \times \sqrt{2} \approx 8.48$, and is the unique shortest path. In taxicab geometry, the green line's length is still 12, making it no shorter than any other path shown.

The Euclidean distance between locations often represents their proximity, although this is only one possibility. There are an infinite number of distances in addition to Euclidean that can support quantitative analysis. For example, "Manhattan" (or "Taxicab") distances where movement is restricted to paths parallel to the axes can be more meaningful than Euclidean distances in urban settings. In addition to distances, other geographic relationships such as connectivity (e.g., the existence or degree of shared borders) and direction can also influence the relationships among entities. It is also possible to compute minimal cost paths across a cost surface; for example, this can represent proximity among locations when travel must occur across rugged terrain.

Types of spatial Analysis

Spatial data comes in many varieties and it is not easy to arrive at a system of classification that is simultaneously exclusive, exhaustive, imaginative, and satisfying. -- G. Upton & B. Fingelton

Spatial Data Analysis

Urban and Regional Studies deal with large tables of spatial data obtained from censuses and surveys. It is necessary to simplify the huge amount of detailed information in order to extract the main trends. Multivariable analysis (or Factor analysis, FA) allows a change of variables, transforming the many variables of the census, usually correlated between themselves, into fewer independent "Factors" or "Principal Components" which are, actually, the eigenvectors of the data correlation matrix weighted by the inverse of their eigenvalues. This change of variables has two main advantages:

1. Since information is concentrated on the first new factors, it is possible to keep only a few of them while losing only a small amount of information; mapping them produces fewer and more significant maps

2. The factors, actually the eigenvectors, are orthogonal by construction, i.e. not correlated. In most cases, the dominant factor (with the largest eigenvalue) is the Social Component, separating rich and poor in the city. Since factors are not-correlated, other smaller processes than social status, which would have remained hidden otherwise, appear on the second, third, ... factors.

Factor analysis depends on measuring distances between observations : the choice of a significant metric is crucial. The Euclidean metric (Principal Component Analysis), the Chi-Square distance (Correspondence Analysis) or the Generalized Mahalanobis distance (Discriminant Analysis) are among the more widely used. More complicated models, using communalities or rotations have been proposed.

Using multivariate methods in spatial analysis began really in the 1950s (although some examples go back to the beginning of the century) and culminated in the 1970s, with the increasing power and accessibility of computers. Already in 1948, in a seminal publication, two sociologists, Bell and Shevky, had shown that most city populations in the USA and in the world could be represented with three independent factors : 1- the « socio-economic status » opposing rich and poor districts and distributed in sectors running along highways from the city center, 2- the « life cycle », i.e. the age structure of households, distributed in concentric circles, and 3- « race and ethnicity », identifying patches of migrants located within the city. In 1961, in a groundbreaking study, British geographers used FA to classify British towns. Brian J Berry, at the University of Chicago, and his students made a wide use of the method, applying it to most important cities in the world and exhibiting common social structures. The use of Factor Analysis in Geography, made so easy by modern computers, has been very wide but not always very wise.

Since the vectors extracted are determined by the data matrix, it is not possible to compare factors obtained from different censuses. A solution consists in fusing together several census matrices in a unique table which, then, may be analyzed. This, however, assumes that the definition of the variables has not changed over time and produces very large tables, difficult to manage. A better solution, proposed by psychometricians, groups the data in a « cubic matrix », with three entries (for instance, locations, variables, time periods). A Three-Way Factor Analysis produces then three groups of factors related by a small cubic « core matrix ». This method, which exhibits data evolution over time, has not been widely used in geography. In Los Angeles, however, it has exhibited the role, traditionally ignored, of Downtown as an organizing center for the whole city during several decades.

Spatial Autocorrelation

Spatial autocorrelation statistics measure and analyze the degree of dependency among observations in a geographic space. Classic spatial autocorrelation statistics include Moran's I, Geary's C, Getis's G and the standard deviational ellipse. These statistics require measuring a spatial weights matrix that reflects the intensity of the geographic relationship between observations in a neighborhood, e.g., the distances between neighbors, the lengths of shared border, or whether they fall into a specified directional class such as "west". Classic spatial autocorrelation statistics compare the spatial weights to the covariance relationship at pairs of locations. Spatial autocorrelation that is more positive than expected from random indicate the clustering of similar values across geographic space, while significant negative spatial autocorrelation indicates that neighboring values are more dissimilar than expected by chance, suggesting a spatial pattern similar to a chess board.

Spatial autocorrelation statistics such as Moran's I and Geary's C are global in the sense that they estimate the overall degree of spatial autocorrelation for a dataset. The possibility of spatial heterogeneity suggests that the estimated degree of autocorrelation may vary significantly across geographic space. Local spatial autocorrelation statistics provide estimates disaggregated to the level of the spatial analysis units, allowing assessment of the dependency relationships across space. G statistics compare neighborhoods to a global average and identify local regions of strong autocorrelation. Local versions of the I and C statistics are also available.

Spatial Stratified Heterogeneity

Spatial stratified heterogeneity, referring to the within-strata variance less than the between strata-variance, is ubiquitous in ecological phenomena, such as ecological zones and many ecological variables. Spatial stratified heterogeneity reflects the essence of nature, implies potential distinct mechanisms by strata, suggests possible determinants of the observed process, allows the representativeness of observations of the earth, and enforces the applicability of statistical inferences. Spatial stratified heterogeneity of an attribute can be measured by geographical detector q-statistic:

$$q = 1 - \frac{1}{N\sigma^2} \sum_{h=1}^{L} N_h \sigma_h^2$$

where a population is partitioned into $h = 1, ..., L$ strata; N stands for the size of the population, σ^2 stands for variance of the attribute. The value of q is within [0, 1], 0 indicates no spatial stratified heterogeneity, 1 indicates perfect spatial stratified heterogeneity. The value of q indicates the percent of the variance of an attribute explained by the stratification. The q follows a noncentral F probability density function.

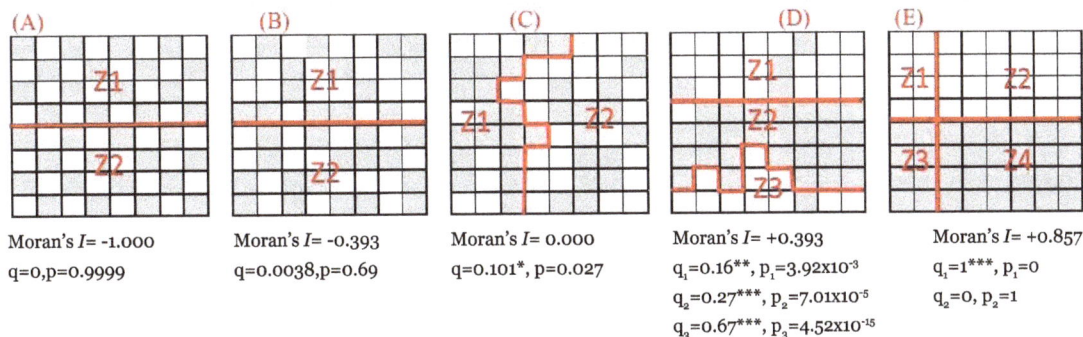

(A)	(B)	(C)	(D)	(E)

Moran's I= -1.000
q=0,p=0.9999

Moran's I= -0.393
q=0.0038,p=0.69

Moran's I= 0.000
q=0.101*, p=0.027

Moran's I= +0.393
q_1=0.16**, p_1=3.92x10^{-3}
q_2=0.27***, p_2=7.01x10^{-5}
q_3=0.67***, p_3=4.52x10^{-15}

Moran's I= +0.857
q_1=1***, p_1=0
q_2=0, p_2=1

A hand map with different spatial patterns. Note: p is the probability of q-statistic; * denotes statistical significant at level 0.05, ** for 0.001, *** for smaller than 10^{-3};(D) subscripts 1, 2, 3 of q and p denotes the strata Z1+Z2 with Z3,Z1 with Z2+Z3, and Z1 and Z2 and Z3 individually, respectively; (E) subscripts 1 and 2 of q and p denotes the strata Z1+Z2 with Z3+Z4,and Z1+Z3 with Z2+Z4, respectively.

Spatial Interpolation

Spatial interpolation methods estimate the variables at unobserved locations in geographic space based on the values at observed locations. Basic methods include inverse distance weighting: this attenuates the variable with decreasing proximity from the observed location. Kriging is a more sophisticated method that interpolates across space according to a spatial lag relationship that has both systematic and random components. This can accommodate a wide range of spatial relationships for the hidden values between observed locations. Kriging provides optimal estimates given the hypothesized lag relationship, and error estimates can be mapped to determine if spatial patterns exist.

Spatial Regression

Spatial regression methods capture spatial dependency in regression analysis, avoiding statistical problems such as unstable parameters and unreliable significance tests, as well as providing information on spatial relationships among the variables involved. Depending on the specific technique, spatial dependency can enter the regression model as relationships between the independent variables and the dependent, between the dependent variables and a spatial lag of itself, or in the error terms. Geographically weighted regression (GWR) is a local version of spatial regression that generates parameters disaggregated by the spatial units of analysis. This allows assessment of the spatial heterogeneity in the estimated relationships between the independent and dependent variables. The use of Bayesian hierarchical modeling in conjunction with Markov Chain Monte Carlo (MCMC) methods have recently shown to be effective in modeling complex relationships using Poisson-Gamma-CAR, Poisson-lognormal-SAR, or Overdispersed logit models. Spatial stochastic processes, such as Gaussian processes are also increasingly being deployed in spatial regression analysis. Model-based versions of GWR, known as spatially varying coefficient models have been applied to conduct Bayesian inference. Spatial stochastic process can become computationally effective and scalable Gaussian process models, such as Gaussian Predictive Processes and Nearest Neighbor Gaussian Processes (NNGP).

Spatial Interaction

Spatial interaction or "gravity models" estimate the flow of people, material or information between locations in geographic space. Factors can include origin propulsive variables such as the number of commuters in residential areas, destination attractiveness variables such as the amount of office space in employment areas, and proximity relationships between the locations measured in terms such as driving distance or travel time. In addition, the topological, or connective, relationships between areas must be identified, particularly considering the often conflicting relationship between distance and topology; for example, two spatially close neighborhoods may not display any significant interaction if they are separated by a highway. After specifying the functional forms of these relationships, the analyst can estimate model parameters using observed flow data and standard estimation techniques such as ordinary least squares or maximum likelihood. Competing destinations versions of spatial interaction models include the proximity among the destinations (or origins) in addition to the origin-destination proximity; this captures the effects of destination (origin) clustering on flows. Computational methods such as artificial neural networks can also estimate spatial interaction relationships among locations and can handle noisy and qualitative data.

Simulation and Modeling

Spatial interaction models are aggregate and top-down: they specify an overall governing relationship for flow between locations. This characteristic is also shared by urban models such as those based on mathematical programming, flows among economic sectors, or bid-rent theory. An alternative modeling perspective is to represent the system at the highest possible level of disaggregation and study the bottom-up emergence of complex patterns and relationships from behavior and interactions at the individual level.

Complex adaptive systems theory as applied to spatial analysis suggests that simple interactions among proximal entities can lead to intricate, persistent and functional spatial entities at aggregate levels. Two fundamentally spatial simulation methods are cellular automata and agent-based modeling. Cellular automata modeling imposes a fixed spatial framework such as grid cells and specifies rules that dictate the state of a cell based on the states of its neighboring cells. As time progresses, spatial patterns emerge as cells change states based on their neighbors; this alters the conditions for future time periods. For example, cells can represent locations in an urban area and their states can be different types of land use. Patterns that can emerge from the simple interactions of local land uses include office districts and urban sprawl. Agent-based modeling uses software entities (agents) that have purposeful behavior (goals) and can react, interact and modify their environment while seeking their objectives. Unlike the cells in cellular automata, simulysts can allow agents to be mobile with respect to space. For example, one could model traffic flow and dynamics using agents representing individual vehicles that try to minimize travel time between specified origins and destinations. While pursuing minimal travel times, the agents must avoid collisions with other vehicles also seeking to minimize their travel times. Cellular automata and agent-based modeling are complementary modeling strategies. They can be integrated into a common geographic automata system where some agents are fixed while others are mobile.

Multiple-point Geostatistics (MPS)

Spatial analysis of a conceptual geological model is the main purpose of any MPS algorithm. The method analyzes the spatial statistics of the geological model, called the training image, and generates realizations of the phenomena that honor those input multiple-point statistics.

A recent MPS algorithm used to accomplish this task is the pattern-based method by Honarkhah. In this method, a distance-based approach is employed to analyze the patterns in the training image. This allows the reproduction of the multiple-point statistics, and the complex geometrical features of the training image. Each output of the MPS algorithm is a realization that represents a random field. Together, several realizations may be used to quantify spatial uncertainty.

One of the recent methods is presented by Tahmasebi et al. uses a cross-correlation function to improve the spatial pattern reproduction. They call their MPS simulation method as the CCSIM algorithm. This method is able to quantify the spatial connectivity, variability and uncertainty. Furthermore, the method is not sensitive to any type of data and is able to simulate both categorical and continuous scenarios. CCSIM algorithm is able to be used for any stationary, non-stationary and multivariate systems and it can provide high quality visual appeal model.,

Geographic Information Science and spatial Analysis

Geographic information systems (GIS) and the underlying geographic information science that advances these technologies have a strong influence on spatial analysis. The increasing ability to capture and handle geographic data means that spatial analysis is occurring within increasingly data-rich environments. Geographic data capture systems include remotely sensed imagery, environmental monitoring systems such as intelligent transportation systems, and location-aware technologies such as mobile devices that can report location in near-real time. GIS provide platforms for managing these data, computing spatial relationships such as distance, connectivity and directional relationships between spatial units, and visualizing both the raw data and spatial analytic results within a cartographic context.

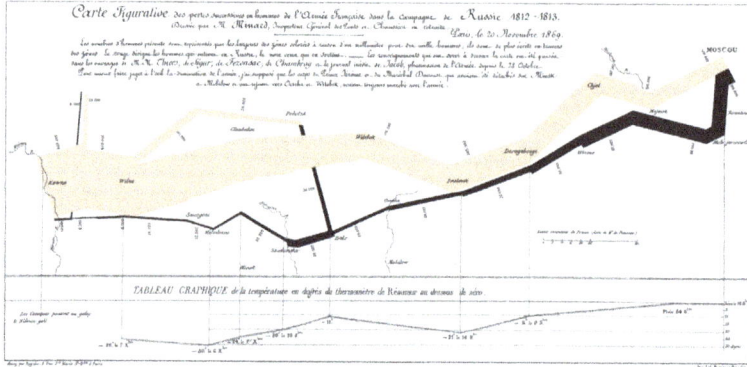

This flow map of Napoleon's ill-fated march on Moscow is an early and celebrated example of geovisualization. It shows the army's direction as it traveled, the places the troops passed through, the size of the army as troops died from hunger and wounds, and the freezing temperatures they experienced.

Geovisualization (GVis) combines scientific visualization with digital cartography to support the exploration and analysis of geographic data and information, including the results of spatial analysis or simulation. GVis leverages the human orientation towards visual information processing in the exploration, analysis and communication of geographic data and information. In contrast with traditional cartography, GVis is typically three- or four-dimensional (the latter including time) and user-interactive.

Geographic knowledge discovery (GKD) is the human-centered process of applying efficient computational tools for exploring massive spatial databases. GKD includes geographic data mining, but also encompasses related activities such as data selection, data cleaning and pre-processing, and interpretation of results. GVis can also serve a central role in the GKD process. GKD is based on the premise that massive databases contain interesting (valid, novel, useful and understandable) patterns that standard analytical techniques cannot find. GKD can serve as a hypothesis-generating process for spatial analysis, producing tentative patterns and relationships that should be confirmed using spatial analytical techniques.

Spatial decision support systems (SDSS) take existing spatial data and use a variety of mathematical models to make projections into the future. This allows urban and regional planners to test intervention decisions prior to implementation.

Vector Operations and Analysis- Single Theme

Nodepoint: Creates a new point theme from the nodes of arcs

Imagine a road theme as shown in Figure (a) having road feature- a line theme where nodes are present at locations where two or more roads meet. At these intersections lie the traffic light poles. Now if one only wants to see the location of traffic lights in the area, he/she can use the Nodepoint to extract the point theme from the nodes of the line theme to represent the location of traffic light poles as shown in Figure (b).

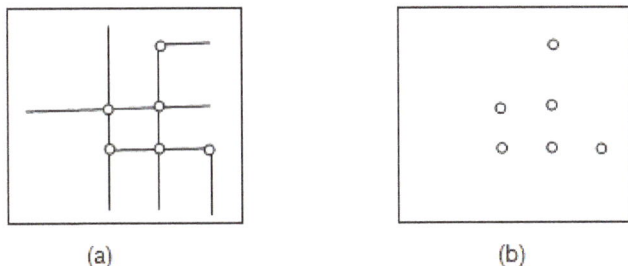

(a)   (b)

Buffer: A buffer is a zone with a width created around a spatial feature and is measured in units of distance from the feature. The generated buffer takes the shape of the feature. In case of a point the buffer is a circle with a radius equal to the buffer distance. In case of a line, it is a band and for a polygon it is a belt of a specific buffer distance from the edge of polygon, surrounding the polygon. The inward buffer for a polygon is called setback (refer Figure (c), the polygon on the right hand side).

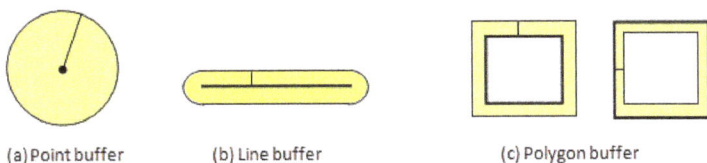

(a) Point buffer   (b) Line buffer   (c) Polygon buffer

Buffering is used for neighborhood analysis which aims to evaluate the characteristics of the area surrounding the spatial feature. Common examples of buffering include the identification of properties within a certain distance of an object, delineation of areas around natural features where human activities are restricted, determination of areas affected by location etc.

Dissolve: It merges the adjacent features if they have the same attribute value and reduces the records from the attribute table. Let us understand it through following examples.

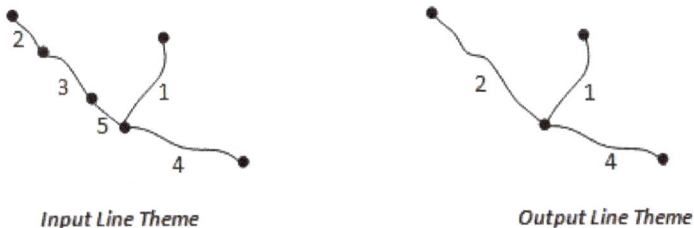

Input Line Theme   Output Line Theme

Type is used as the dissolve item.

River Attribute Table		
River#	RiverID	Type
1	1	2
2	2	1
3	3	1
4	4	2
5	5	1

New_River Attribute Table		
New_River#	New_RiverID	Type
1	1	2
2	2	1
3	4	2

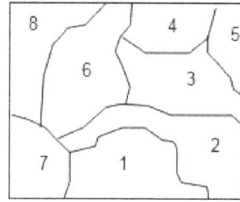

Input Polygon theme

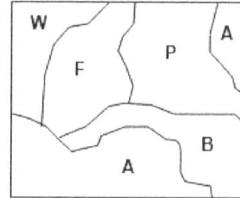

Output Polygon theme

Here, Landuse type is used as the dissolve item

Landuse Attribute table		
LUPoly#	L_ID	LanduseType
1	1	Agriculture (A)
2	2	Barren (B)
3	3	Pasture (P)
4	4	Pasture (P)
5	5	Agriculture (A)
6	6	Forest (F)
7	7	Agriculture (A)
8	8	Water (W)

New_Landuse Attribute table		
New_LUPoly#	New_LID	LanduseType
1	1	Agriculture (A)
2	2	Barren (B)
3	3	Pasture (P)
4	6	Forest (F)
5	8	Water (W)

Vector Operations and Analysis- Multiple Theme

These operations work on the layers at a time rather than selective spatial features. The layers which are used must be topologically structured so as to get a correct, topologically structured output.

Clip is used to subset a point, line or a polygon theme using another polygon theme as the boundary of the area of interest.

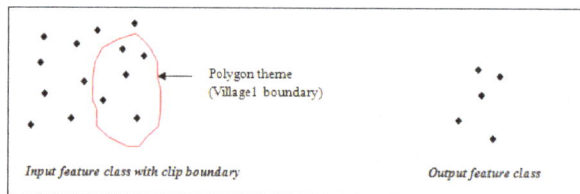

Polygon theme
(Village1 boundary)

Input feature class with clip boundary

Output feature class

In the illustration above, the input, point feature shows the location of drinking water wells in three villages. To know how many wells fall in village1, the input feature class is clipped using the boundary of the village1. The output feature class shows that five wells are present in village1.

Split causes the input features to form subset of multiple output feature classes. The split field's unique values form the names of the output feature classes.

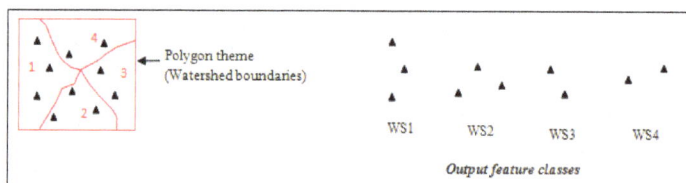

Polygon theme
(Watershed boundaries)

WS1   WS2   WS3   WS4

Output feature classes

In the illustration above, a point theme of wells is split using the polygon theme of watershed boundaries. The output of this operation contains multiple feature classes which are named on the unique value of watershed boundaries (in this case, the unique value is the watershed number WS1, WS2 etc.). Each output class represents the number of wells present in a particular watershed i.e. WS1 or watershed 1 has three wells. Similarly, WS2, WS3 and WS4 have 3, 2, and 2 wells respectively.

Overlay Operations

Union creates a new theme by overlaying two polygon themes. It is same as 'or' Boolean operator. The output theme contains the combined polygons and attributes of both themes. Only polygon themes can be combined using union.

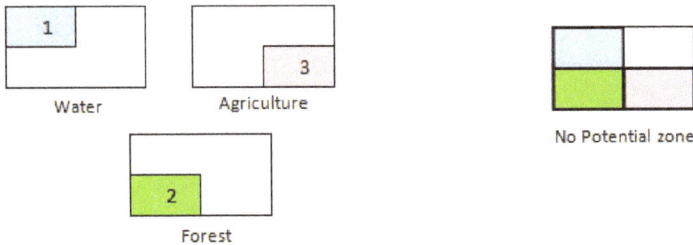

Let's say we are interested in knowing no potential zone for urban development. It is clear that no construction can be done on a waterbody or land covered by agriculture or forest. So, we can say union of areas under water, agriculture and forest would present us the area having no potential for urban development.

Intersect creates a new theme by overlaying a point, line or polygon theme with an intersecting polygon theme. It is same as 'and' Boolean operator. The output theme contains only the feature inside the intersecting polygons.

From the same example given above, if we try to know the area having potential for urban development we need to intersect the polygon themes to get a common area which is not under water, agriculture or forest.

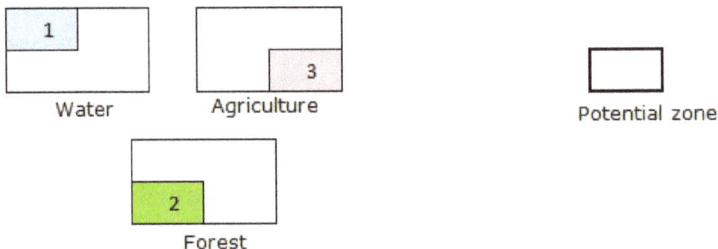

Identity creates a new theme by overlaying a point, line, or polygon theme with an intersecting polygon theme. The output theme contains all the original point, line, or polygons as well as the attributes transferred by the intersecting polygon theme.

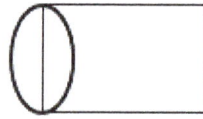

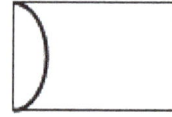

Erase removes polygons in a theme from the area covered by polygons of another theme. The output feature class only contains those features of the input polygon theme that fall outside the polygons of the second theme.

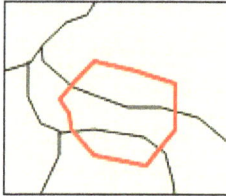

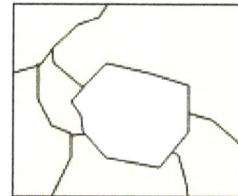

Network Analysis

It is a type of line analysis which involves set of interconnected lines. Railways, highways, transportation routes, rivers etc are examples of networks. Network analysis can be used for the following:

Address Geocoding

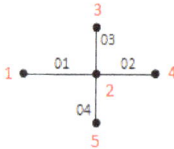

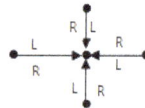

Arc#	From_node	To_node
01	1	2
02	4	2
03	3	2
04	5	2

Arc	Length	Left_from	Left_to	Right_from	Right_to	Prefix	Name	Type	Suffix
01	2000	100	1300	101	1299	West	Main	Street	
02	1800	2801	1401	2800	1400	East	Main	Street	
03	1000	200	248	201	249		First	Avenue	North
04	800	299	251	300	250		First	Avenue	South

It is the process of estimating the locations of addresses in GIS coordinate system. It requires a table of addresses and theme that contains attributes that can be used to match to the table of addresses.

Imagine that the fire department is reported about a fire in a building at 1000 West Main Street. To estimate the location, GIS determines the arc by matching its name, type and suffix. Once the arc is determined the address can be estimated using linear interpolation. The arc corresponding to West Main Street is the Arc 01.

The address is an even number and lies on the left side of the arc. The left side has addresses ranging from 100 to 1300 (range is 1200). The length of the arc is 2000 meters. The address of 1000 can be geocoded as

$$\frac{(\text{Max. address} - \text{address})}{\text{Range}} \times \text{Arc length}$$

$$= \frac{(1300 - 1000)}{1200} \times 2000$$

$$= 500 \text{ metres along the arc}$$

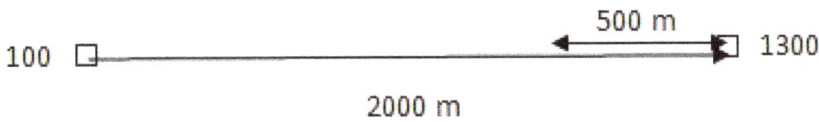

Optimal Routing

Optimal routing is the process of finding out the best route to go from one location to another location. The most common path finding algorithm is Dijkstra algorithm which was published by E.W. Dijkstra in 1956. It is a graph search algorithm that provides the shortest path for a single source shortest path problem.

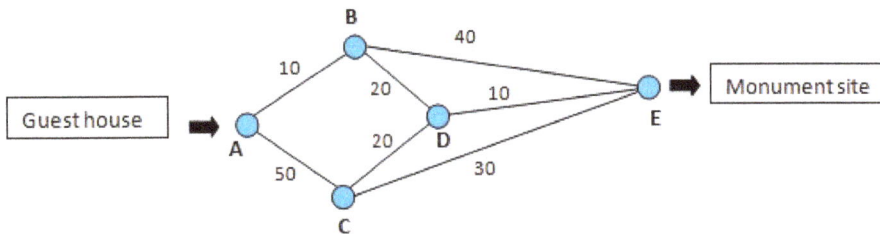

Imagine we have the following network of roads. We want to find the quickest path to get from the guest house to monument site. The variation in time travelling along different roads depends upon the traffic conditions.

We build two tables, one for the nodes that have been already processed and the other for the adjacent nodes which are to be processed. We begin with Node A as follows:

Processed Nodes		
Node	Cumulative cost	Previous node
A	0	none

Adjacent Nodes		
Node	Cumulative cost	Previous node
B	10	A
C	50	A

We pick the adjacent node with least cumulative cost, and place it in the processed node table.

Processed Nodes		
Node	Cumulative cost	Previous node
A	0	none
B	10	A

Adjacent Nodes		
Node	Cumulative cost	Previous node
C	50	A
D	30	B
E	50	B

Processed Nodes		
Node	Cumulative cost	Previous node
A	0	none
B	10	A
D	30	B

Adjacent Nodes		
Node	Cumulative cost	Previous node
C	50	A
E	50	B
C	50	D
E	40	D

Processed Nodes		
Node	Cumulative cost	Previous node
A	0	none
B	10	A
D	30	B
E	40	D

Node E is the destination so we stop here. The quickest route to reach node E takes 40 minutes and it is Node A $\rightarrow$ Node B $\rightarrow$ Node C $\rightarrow$ Node D

Finding Closest Facilities

Sometimes we try to find out a point closest to a given location. The point is called a facility and the given location is called an event location. Finding which flat would be near to the working place, which fire station has the best response time to a report to the fire location, which houses are close to the schools are examples of optimal routing for closest facilities.

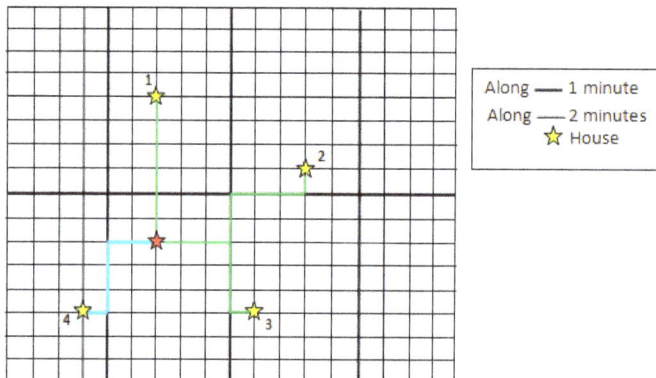

In the above illustration, imagine that the office (red star) is located at some address and we want to know the closest place around the office where an employee can live. The address of the office is geocoded to a street location and then optimal path can be computed from each house to the office.

House	Estimated Travel Time (min)
1	12
2	13
3	11
4	9

House 4 is closest to the office as the travel time is least for it.

Geocoding

Geocoding is the computational process of transforming a postal address description to a location on the Earth's surface (spatial representation in numerical coordinates). Reverse geocoding, on the other hand, converts the inputted geographic coordinates to a description of a location, usually the name of a place or a postal address. Geocoding relies on a computer representation of the street network. Geocoding is sometimes used for conversion from ZIP codes or postal codes to coordinates, occasionally for the conversion of parcel identifiers to centroid coordinates.

Geocoding (*verb*): The act of transforming an address text into a valid spatial representation.

Geocoder (*noun*): A piece of software or a (web) service that implements a geocoding process i.e. a set of interrelated components in the form of operations, algorithms, and data sources that work together to produce a spatial representation for descriptive locational references.

Geocode (*noun*): A spatial representation of a descriptive locational reference.

The geographic coordinates representing locations often vary greatly in positional accuracy. Examples include building centroids, land parcels, street addresses, postal code centroids (e.g. ZIP codes, CEDEX), and Administrative Boundary Centroids.

History

Geocoding — a subset of Geographic Information System (GIS) spatial analysis — has been a subject of interest since the early 1960s.

1960s

In 1960, the first operational GIS — named the Canada Geographic Information System (CGIS) — was invented by Dr. Roger Tomlinson, who has since been acknowledged as the father of GIS. The CGIS was used to store and analyze data collected for the Canada Land Inventory, which mapped information about agriculture, wildlife, and forestry at a scale of 1:50,000, in order to regulate land capability for rural Canada. However, the CGIS lasted until the 1990s and was never available commercially.

On July 1, 1963, five-digit ZIP codes were introduced nationwide by the United States Post Office Department (USPOD). In 1983, nine-digit ZIP+4 codes were brought about as an extra identifier in more accurately locating addresses.

In 1964, the Harvard Laboratory for Computer Graphics and Spatial Analysis developed groundbreaking software code — e.g. GRID, and SYMAP — all of which were sources for commercial development of GIS.

In 1967, a team at the Census Bureau — including the mathematician James Corbett and Donald Cooke — invented Dual Independent Map Encoding (DIME) — the first modern vector mapping model — which ciphered address ranges into street network files and incorporated the "percent along" geocoding algorithm. Still in use by platforms such as Google Maps and MapQuest, the "percent along" algorithm denotes where a matched address is located along a reference feature as a percentage of the reference feature's total length. DIME was intended for the use of the United States Census Bureau, and it involved accurately mapping block faces, digitizing nodes representing street intersections, and forming spatial relationships. New Haven, Connecticut was the first city on Earth with a geocodable streets network database.

1980s

In the late 1970s, two main public domain geocoding platforms were in development: GRASS GIS and MOSS. The early 1980s saw the rise of many more commercial vendors of geocoding software, namely Intergraph, ESRI, CARIS, ERDAS, and MapInfo Corporation. These platforms merged the 1960s approach of separating spatial information with the approach of organizing this spatial information into database structures.

In 1986, Mapping Display and Analysis System (MIDAS) became the first desktop geocoding software, designed for the DOS operating system. Geocoding was elevated from the research department into the business world with the acquisition of MIDAS by MapInfo. MapInfo has since been acquired by Pitney Bowes, and has pioneered in merging geocoding with business intelligence; allowing location intelligence to provide solutions for the public and private sectors.

1990s

The end of the 20th century had seen geocoding become more user-oriented, especially via open-source GIS software. Mapping applications and geospatial data had become more accessible over the Internet.

Because the mail-out/mail-back technique was so successful in the 1980 Census, the U.S. Bureau of Census was able to put together a large geospatial database, using interpolated street geocoding. This database — along with the Census' nationwide coverage of households — allowed for the birth of TIGER (Topologically Integrated Geographic Encoding and Referencing).

Containing address ranges instead of individual addresses, TIGER has since been implemented in nearly all geocoding software platforms used today. By the end of the 1990 Census, TIGER "contained a latitude/longitude-coordinate for more than 30 million feature intersections and endpoints and nearly 145 million feature 'shape' points that defined the more than 42 million feature segments that outlined more than 12 million polygons."

TIGER was the breakthrough for "big data" geospatial solutions.

2000s

The early 2000s saw the rise of Coding Accuracy Support System (CASS) address standardization. The CASS certification is offered to all software vendors and advertising mailers who want the United States Postal Services (USPS) to assess the quality of their address-standardizing software. The annually renewed CASS certification is based on delivery point codes, ZIP codes, and ZIP+4 codes. Adoption of a CASS certified software by software vendors allows them to receive discounts in bulk mailing and shipping costs. They can benefit from increased accuracy and efficiency in those bulk mailings, after having a certified database. In the early 2000s, geocoding platforms were also able to support multiple datasets.

In 2003, geocoding platforms were capable of merging postal codes with street data, updated monthly. This process became known as "conflation".

Beginning in 2005, geocoding platforms included parcel-centroid geocoding. Parcel-centroid geocoding allowed for a lot of precision in geocoding an address. For example, parcel-centroid allowed a geocoder to determine the centroid of a specific building or lot of land. Platforms were now also able to determine the elevation of specific parcels.

2005 also saw the introduction of the Assessor's Parcel Number (APN). A jurisdiction's tax assessor was able to assign this number to parcels of real estate. This allowed for proper identification and record-keeping. An APN is important for geocoding an area which is covered by a gas or oil lease, and indexing property tax information provided to the public.

In 2006, Reverse Geocoding and reverse APN lookup were introduced to geocoding platforms. This involved geocoding a numerical point location — with a longitude and latitude — to a textual, readable address.

2008 and 2009 saw the growth of interactive, user-oriented geocoding platforms — namely MapQuest, Google Maps, Bing Maps, and Global Positioning Systems (GPS). These platforms were made even more accessible to the public with the simultaneous growth of the mobile industry, specifically smartphones.

2010s

This current decade has seen vendors fully supporting geocoding and reverse geocoding globally. Cloud-based geocoding application programming interface (API) and on-premise geocoding has allowed for a greater match rate, greater precision, and greater speed. There is now a popularity in the idea of geocoding being able to influence business decisions. This is the integration between the geocoding process and business intelligence.

The future of geocoding also involves three-dimensional geocoding, indoor geocoding, and multiple language returns for the geocoding platforms.

Geocoding Process

Geocoding is a task which involves multiple datasets and processes, all of which work together. A geocoder is made of two important components: a reference dataset and the geocoding algorithm. Each of these components are made up of sub-operations and sub-components. Without understanding how these geocoding processes work, it is difficult to make informed business decisions based on geocoding.

Input Data

Input data are the descriptive, textual information (address or building name) which the user wants to turn into numerical, spatial data (latitude and longitude) — through the process of geocoding.

Classification of Input Data

Input data is classified into two categories: relative input data and absolute input data.

Relative Input Data

Relative input data are the textual descriptions of a location which, alone, cannot output a spatial representation of that location. Such data outputs a relative geocode, which is dependent and geographically relative of other reference locations. An example of a relative geocode is address-interpolation using areal units or line vectors. "Across the street from the Empire State Building" is an example of a relative input data. The location being sought cannot be determined without identifying the Empire State Building. Geocoding platforms often do not support such relative locations, but advances are being made in this direction.

Absolute Input Data

Absolute input data are the textual descriptions of a location which, alone, can output a spatial representation of that location. This data type outputs an absolute known location independently of other locations. For example, USPS ZIP codes; USPS ZIP+4 codes; complete and partial postal addresses; USPS PO boxes; rural routes; cities; counties; intersections; and named places can all be referenced in a data source absolutely.

When there is a lot of variability in the way addresses can be represented — such as too much input data or too little input data — geocoders use address normalization and address standardization in order to resolve this problem.

Processing of Input Data

Address interpolation

A simple method of geocoding is address interpolation. This method makes use of data from a street geographic information system where the street network is already mapped within the geographic coordinate space. Each street segment is attributed with address ranges (e.g. house numbers from one segment to the next). Geocoding takes an address, matches it to a street and specific segment (such as a block, in towns that use the "block" convention). Geocoding then interpolates the position of the address, within the range along the segment.

Example

Take for example: *742 Evergreen Terrace*

Let's say that this segment (for instance, a block) of Evergreen Terrace runs from 700 to 799. Even-numbered addresses fall on the east side of Evergreen Terrace, with odd-numbered addresses on the west side of the street. 742 Evergreen Terrace would (probably) be located slightly less than halfway up the block, on the east side of the street. A point would be mapped at that location along the street, perhaps offset a distance to the east of the street centerline.

Complicating Factors

However, this process is not always as straightforward as in this example. Difficulties arise when

- distinguishing between ambiguous addresses such as 742 Evergreen Terrace and 742 W Evergreen Terrace.

- attempting to geocode new addresses for a street that is not yet added to the geographic information system database.

While there might be 742 Evergreen Terrace in Springfield, there might also be a 742 Evergreen Terrace in Shelbyville. Asking for the city name (and state, province, country, etc. as needed) can solve this problem. Boston, Massachusetts has multiple "100 Washington Street" locations because several cities have been annexed without changing street names, thus requiring use of unique postal codes or district names for disambiguation. Geocoding accuracy can be greatly improved by first utilizing good address verification practices. Address verification will confirm the existence of the address and will eliminate ambiguities. Once the valid address is determined, it is very easy to geocode and determine the latitude/longitude coordinates. Finally, several caveats on using interpolation:

- The typical attribution of a street segment assumes that all even numbered parcels are on one side of the segment, and all odd numbered parcels are on the other. This is often not true in real life.

- Interpolation assumes that the given parcels are evenly distributed along the length of the segment. This is almost never true in real life; it is not uncommon for a geocoded address to be off by several thousand feet.

- Interpolation also assumes that the street is straight. If a street is curved then the geocoded location will not necessarily fit the physical location of the address.

- Segment Information (esp. from sources such as TIGER) includes a maximum upper bound for addresses and is interpolated as though the full address range is used. For example, a segment (block) might have a listed range of 100-199, but the last address at the end of the block is 110. In this case, address 110 would be geocoded to 10% of the distance down the segment rather than near the end.

- Most interpolation implementations will produce a point as their resulting address location. In reality, the physical address is distributed along the length of the segment, i.e. consider geocoding the address of a shopping mall - the physical lot may run a distance along the street segment (or could be thought of as a two-dimensional space-filling polygon which may front on several different streets — or worse, for cities with multi-level streets, a three-dimensional shape that meets different streets at several different levels) but the interpolation treats it as a singularity.

A very common error is to believe the accuracy ratings of a given map's geocodable attributes. Such accuracy currently touted by most vendors has no bearing on an address being attributed to the correct segment, being attributed to the correct side of the segment, nor resulting in an accurate position along that correct segment. With the geocoding process used for U.S. Census TIGER datasets, 5-7.5% of the addresses may be allocated to a different census tract, while a study of Australia's TIGER-like system found that 50% of the geocoded points were mapped to the wrong property parcel. The accuracy of geocoded data can also have a bearing on the quality of research that can be done using this data. One study by a group of Iowa researchers found that the common method of geocoding using TIGER datasets as described above, can cause a loss of as much as 40% of the power of a statistical analysis. An alternative is to use orthophoto or image coded data such as the Address Point data from Ordnance Survey in the UK, but such datasets are generally expensive. Because of this, it is quite important to avoid using interpolated results except for non-critical applications, such as pizza delivery. Interpolated geocoding is usually not appropriate for making authoritative decisions, for example if life safety will be affected by that decision. Emergency services, for example, do not make an authoritative decision based on their interpolations; an ambulance or fire truck will always be dispatched regardless of what the map says.

Other Techniques

In rural areas or other places lacking high quality street network data and addressing, GPS is useful for mapping a location. For traffic accidents, geocoding to a street intersection or midpoint along a street centerline is a suitable technique. Most highways in developed countries have mile markers to aid in emergency response, maintenance, and navigation. It is also possible to use a combination of these geocoding techniques — using a particular technique for certain cases and situations and other techniques for other cases. In contrast to geocoding of structured postal address records, toponym resolution maps place names in unstructured document collections to their corresponding spatial footprints.

Research

Research has introduced a new approach to the control and knowledge aspects of geocoding, by using an agent-based paradigm. In addition to the new paradigm for geocoding, additional correction techniques and control algorithms have been developed. The approach represents the geographic elements commonly found in addresses as individual agents. This provides a commonality and duality to control and geographic representation. In addition to scientific publication, the new approach and subsequent prototype gained national media coverage in Australia. The research was conducted at Curtin University in Perth, Western Australia.

Uses

Geocoded locations are useful in many GIS analysis, cartography, decision making workflow, transaction mash-up, or injected into larger business processes. On the web, geocoding is used in services like routing and local search. Geocoding, along with GPS provides location data for geotagging media, such as photographs or RSS items.

Privacy Concerns

The proliferation and ease of access to geocoding (and reverse-geocoding) services raises privacy concerns. For example, in mapping crime incidents, law enforcement agencies aim to balance the privacy rights of victims and offenders, with the public's right to know. Law enforcement agencies have experimented with alternative geocoding techniques that allow them to mask a portion of the locational detail (e.g., address specifics that would lead to identifying a victim or offender). As well, in providing online crime mapping to the public, they also place disclaimers regarding the locational accuracy of points on the map, acknowledging these location masking techniques, and impose terms of use for the information.

Raster Data Spatial Analysis

Local Operations

Local functions process a grid on a cell-by-cell basis, that is, the output value of each cell depends on the values of corresponding cells in the rasters input for the analysis. The following are the examples of the local operations.

Arithmetic Operation

Grids can undergo a range of arithmetic operations such as addition, subtraction, multiplication, and division. If the data in grids (operands) is in the form of integer then the data in the resultant grid after any mathematical operation would also be integer.

Only in one case, when any integer is divided by zero the corresponding resultant cell will be undefined and are assigned to no data. No data cells always remain No data in arithmetic operations.

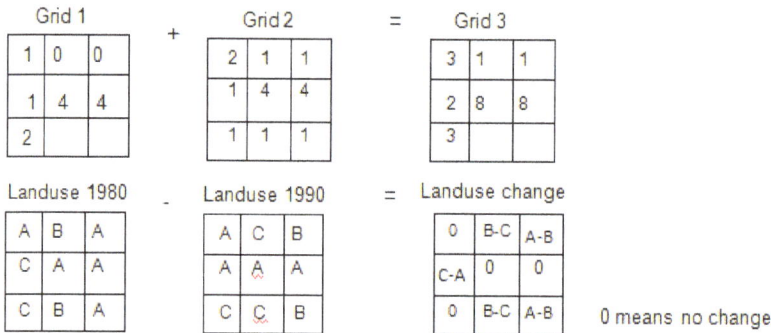

Grid 1
1	0	0
1	4	4
2		

+

Grid 2
2	1	1
1	4	4
1	1	1

=

Grid 3
3	1	1
2	8	8
3		

Landuse 1980
A	B	A
C	A	A
C	B	A

−

Landuse 1990
A	C	B
A	A	A
C	C	B

=

Landuse change
0	B-C	A-B
C-A	0	0
0	B-C	A-B

0 means no change

SELECT returns original cell value if the cell meets the logical expression criteria otherwise the output cell is assigned no data.

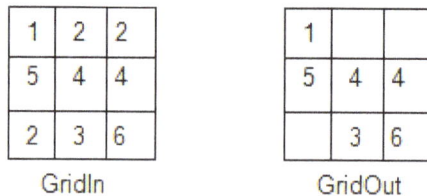

1	2	2
5	4	4
2	3	6

GridIn

1		
5	4	4
	3	6

GridOut

GridOut = SELECT (GridIn, "value ≠ 2")

TEST performs a Boolean evaluation of the input raster using a logical expression. If the cell meets the logical expression criteria it assigns the output cell a value of 1 otherwise the output cell is assigned 0.

1	2	5
4	4	4
2	3	6

GridIn

1	1	0
0	0	0
1	1	0

GridOut

GridOut = TEST (GridIn, "value ≤ 3")

5	3	6
2	4	4
2	1	6

GridIn

1	0	1
0	1	1
0	0	1

GridOut

GridOut = CON (GridIn >3, 1, 0)

CON performs a conditional evaluation on each cell of the input raster. It tests for a user specified logical expression and return user specified values.

Focal Operations (Neighborhood Analysis)

The value of a cell in the output raster depends upon the value of the corresponding cells and their neighboring cells in the input rasters. The neighborhood for a cell is generally taken as a 3×3 matrix (window) in which the cell itself occupies the centre and is surrounded by the others eight cells. With each cell in the input getting processed, the neighborhood window keeps moving.

Spatial Aggregation

It is the process of reducing the number of cells in the raster layer to achieve generalization. It is not a compression technique because the same geographic space is represented using small number of cells of coarse spatial resolution. The choice of neighborhood window size results in different outputs. A large window size will result in higher aggregation level which implies a greater loss of details. A user can use one of the following aggregating methods:

- Averaging method: It compute average value of the cells over the window and is used as the value of the aggregated cell

- Central cell method: Value of the cell at the centre of the window is the value of the aggregated cell.

- Median method: It computes the median value of all the cells over the window and uses it as the value of the aggregated cell.

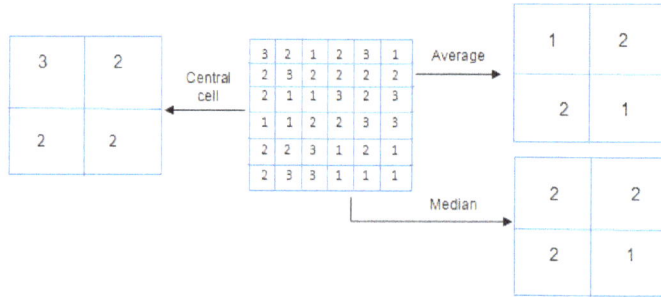

Filtering

Filtering is used to enhance an image. The value in each cell of the raster image indicates the degree of brightness at that point. The change in brightness value per unit distance for any part of the image gives rise to an image characteristic called spatial frequency. If the change is small, the area is of low frequency and if the brightness values change rapidly over small distance the area is of high frequency. The method of filtering uses a filter window or kernel (generally a 3× 3 matrix) which is passed over the whole image. On passing the kernel over the image, the coefficients in the kernels are multiplied by the value of the corresponding cells and the average value is assigned to the cell at the centre of the kernel.

Filtering suppresses noise in the image and highlight specific characteristics.

Original Image

Smooth image obtained after Low pass filtering

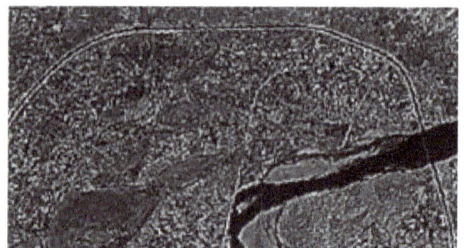

Sharpened image obtained after high pass filtering

This neighborhood operation is used for DEMs. The angle of slope can be calculated as:

$$\tan a = \frac{vd}{hd}$$

$$a = \tan^{-1} \frac{vd}{hd}$$

Where vd is the difference in height between two points and hd is the horizontal distance between the two points for which slope is to be determined. The illustration below described the method of slope calculation.

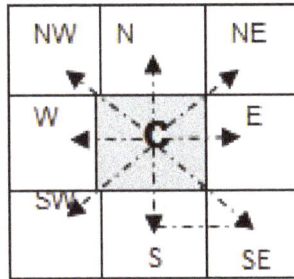

In a 3×3 window of the raster layer, the centre cell is surrounded by 8 cells and each can be given a direction. We can calculate the slope angle of the centre cell for each of these directions. The horizontal distance between the centre cell and the cells in N, E, W and S is the cell size of the raster data and the vertical distance is the difference of height between the centre cell and each of the four cells. In case of cells in NW, NE, SW and SE, the horizontal distance between the centre cell and these cells can be calculated as per Pythagoras formula:

$$hd = \sqrt{(S-C)^2 + (S-SE)^2}$$

The values of vd and hd are substituted in the equation to determine the angle of slope.

Zonal Operations

Zonal functions process grid in such a manner that cells of same zone are analyzed together. A zone may or may not be contiguous. The output value for each location depends upon the value of cell at that location and the association the location has within a zone.

Zonal geometry: The zonal geometric functions return geometric information about each zone in a grid. Following geometries can be calculated using it.

a. Area: For each zone in the input raster, Zonal area calculates the area and assign it to each cell of the zone on the output raster. The area is calculated by the number of cells that comprise the zone multiplied by the current cell size.

1	2	2	3
4	1	1	3
4	5	5	2
1	1	5	2

5	4	4	2
2	5	5	2
2	3	3	4
5	5	3	4

 # Cell size is 1

b. Perimeter: Zonalperimeter determines the perimeter of each zone on the input raster and assigns it to each cell in the zone on an output raster.

1	2	2	3
4	1	1	3
4	5	5	2
1	1	5	2

16	12	12	6
6	16	16	6
6	8	8	12
16	16	8	12

 # Cell size is 1

c. Centroid: Zonal centroid approximates the geometry of each zone by creating an ellipse fixed at the centroid of each zonal spatial shape. The area of each ellipse is equal to the area of the zone it represents.

1	1	2	2	5	5
1	2	2	2	5	5
	2	2	2	4	4
3	3	4	4	4	3
3	6	6	6	4	3
3	3	6	6	6	3

1					
		2		5	
			4		
		3			
			6		

Zonal statistics: It calculates statistics for each zone of the input dataset. The zonal statistical functions return a statistical measure of the values of each zone. The measure can be the mean, median, majority, standard deviation, sum, minimum, maximum, or range of the input values.

Global Operations

The value of each cell in the output raster is a function of the entire grid. Following are few of the global functions:

REGIONGROUP: It aggregates the cells with same values into groups. It starts at upper left corner cell and proceed left to right assigning group numbers based on cells that touch and have the same cell values.

4	3	4	3
3	3	4	3
4	4	4	3
4	4	3	3

GridIn

1	2	3	4
2	2	3	4
3	3	3	4
3	3	4	4

GridOut

Value attribute table		
Value	Count	Link
1	1	4
2	3	3
3	7	4
4	5	3

The value attribute table contains link along with the values and their count. Link is the original value of the cells before the grouping occurred.

EUCDISTANCE: The Euclidean distance grid identifies the distance from each cell to the closest source cell.

1	1		
	1		
		2	

0	0	1	2
1	0	1	2
1.4	1	1	1.4
2	1	0	1

The output values for the Euclidean grid are floating-point distance values. If a cell is at an equal distance from two or more sources, it is assigned to the source that is first encountered in the scanning process (The process of scanning starts at the upper left corner and moves from left-to-right, top-to-bottom).

Other global functions include Costdistance, Costpath, Eucdirection, Slice etc.

References

- Olivares, Miriam. "Geographic Information Systems at Yale: Geocoding Resources". guides.library.yale.edu. Retrieved 2016-06-22

- Banerjee, Sudipto; Gelfand, Alan E.; Finley, Andrew O.; Sang, Huiyan (2008). "Gaussian predictive process models for large spatial datasets". Journal of the Royal Statistical Society Series B. 70 (4): 825–848. doi:10.1111/j.1467-9868.2008.00663.x

- (eds.), Wolfgang Kresse, David M. Danko (2010). Springer handbook of geographic information (1. ed.). Berlin: Springer. pp. 82–83. ISBN 9783540726807. CS1 maint: Extra text: authors list (link)

- "Spatially enabling the data: What is geocoding?". National Criminal Justice Reference Service. Retrieved 2016-06-22

- Datta, Abhirup; Banerjee, Sudipto; Finley, Andrew O.; Gelfand, Alan E. (2016). "Hierarchical Nearest Neighbor Gaussian Process Models for Large Geostatistical Datasets". Journal of the American Statistical Association. doi:10.1080/01621459.2015.1044091

- Jennifer Foreshew (24 November 2009). "Difficult addresses no problem for IntelliGeoLocator". The Australian. Retrieved 9 May 2011

- Banerjee, Sudipto; Carlin, Bradley P.; Gelfand, Alan E. (2014), Hierarchical Modeling and Analysis for Spatial Data, Second Edition, Monographs on Statistics and Applied Probability (2nd ed.), Chapman and Hall/CRC, ISBN 9781439819173

- Ratcliffe, Jerry H. (2001). "On the accuracy of TIGER-type geocoded address data in relation to cadastral and census areal units" (PDF). International Journal of Geographical Information Science. 15 (5). Archived from the original (PDF) on 23 June 2006

- Department of Education, Western Australia (April 2011). "X marks the spot". School Matters. Retrieved 9 May 2011

Permissions

All chapters in this book are published with permission under the Creative Commons Attribution Share Alike License or equivalent. Every chapter published in this book has been scrutinized by our experts. Their significance has been extensively debated. The topics covered herein carry significant information for a comprehensive understanding. They may even be implemented as practical applications or may be referred to as a beginning point for further studies.

We would like to thank the editorial team for lending their expertise to make the book truly unique. They have played a crucial role in the development of this book. Without their invaluable contributions this book wouldn't have been possible. They have made vital efforts to compile up to date information on the varied aspects of this subject to make this book a valuable addition to the collection of many professionals and students.

This book was conceptualized with the vision of imparting up-to-date and integrated information in this field. To ensure the same, a matchless editorial board was set up. Every individual on the board went through rigorous rounds of assessment to prove their worth. After which they invested a large part of their time researching and compiling the most relevant data for our readers.

The editorial board has been involved in producing this book since its inception. They have spent rigorous hours researching and exploring the diverse topics which have resulted in the successful publishing of this book. They have passed on their knowledge of decades through this book. To expedite this challenging task, the publisher supported the team at every step. A small team of assistant editors was also appointed to further simplify the editing procedure and attain best results for the readers.

Apart from the editorial board, the designing team has also invested a significant amount of their time in understanding the subject and creating the most relevant covers. They scrutinized every image to scout for the most suitable representation of the subject and create an appropriate cover for the book.

The publishing team has been an ardent support to the editorial, designing and production team. Their endless efforts to recruit the best for this project, has resulted in the accomplishment of this book. They are a veteran in the field of academics and their pool of knowledge is as vast as their experience in printing. Their expertise and guidance has proved useful at every step. Their uncompromising quality standards have made this book an exceptional effort. Their encouragement from time to time has been an inspiration for everyone.

The publisher and the editorial board hope that this book will prove to be a valuable piece of knowledge for students, practitioners and scholars across the globe.

Index

www.ingramcontent.com/pod-product-compliance
Lightning Source LLC
Chambersburg PA
CBHW062003190326

41458CB00009B/2954